Seeing like a Rover

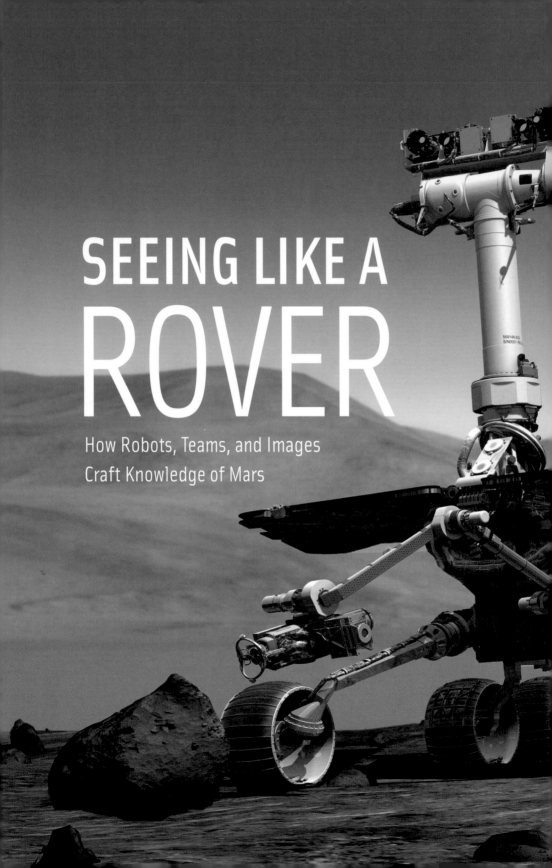

SEEING LIKE A
ROVER

How Robots, Teams, and Images
Craft Knowledge of Mars

JANET VERTESI

The University of Chicago Press
Chicago and London

Janet Vertesi is assistant professor at Princeton University.

The University of Chicago Press, Chicago 60637
The University of Chicago Press, Ltd., London

24 23 22 21 20 19 18 17 16 15   1 2 3 4 5

ISBN-13: 978-0-226-15596-8   (cloth)
ISBN-13: 978-0-226-15601-9   (e-book)

DOI: 10.7208/chicago/9780226156019.001.0001

Vertesi, Janet, author.
    Seeing like a Rover : how robots, teams, and images craft knowledge of Mars / Janet Vertesi.
        pages ; cm
    Includes bibliographical references and index.
    ISBN 978-0-226-15596-8 (cloth : alk. paper) — ISBN 978-0-226-15601-9 (e-book)
1. Mars Exploration Rover Mission (U.S.)—Data processing. 2. Mars (Planet)—Exploration—Data processing. 3. Roving vehicles (Astronautics)—Automatic control—Data processing. 4. Image processing—Digital techniques. I. Title.
    TL799.M3V47 2015
    523.43072'3—dc23
                                                                2014029027

This publication is made possible in part from the Barr Ferree Foundation Fund for Publications, Princeton University.

♾ This paper meets the requirements of ANSI/NISO Z39.48-1992 (Permanence of Paper).

For Catherine and Les

After you've worked with the team for a while, you kind of learn to see like a Rover.

—Jude, Mars Exploration Rover team member

CONTENTS

ACKNOWLEDGMENTS

This book had its roots in the extraordinary Science & Technology Studies Department at Cornell University, where I spent five wonderful years and incurred many debts to my advisers and fellow students. Michael Lynch's tireless and exemplary guidance, support, and intellectual engagement were nothing short of extraordinary. Trevor Pinch and Phoebe Sengers each placed their unique stamp on this project and on my thinking as a scholar, for which I will always be grateful. Meanwhile, Jofish Kaye, Shay David, Lisa Onaga, Nicole Nelson, and Katie Proctor helped me think through tough problems while brightening even the grayest of Ithaca days. An NSF grant gave me the ability to travel to many Rover sites, and a National Aeronautics and Space Administration History Office/History of Science Society Fellowship in the History of Space Sciences helped me conduct the archival and oral history research essential to the book. A Mellon Fellowship at the Society for Humanities at Cornell provided an excellent place to think and write, while an extended visit to NASA Ames Research Center's History and Intelligent Systems divisions and to Stanford University's STS program provided an environment for finishing touches.

I have also benefited from two postdoctoral opportunities that helped this project develop in exciting directions. Thanks to a grant from the NSF in virtual organizations as sociotechnical systems, I spent eighteen months at the Informatics Department of the University of California, Irvine, under the auspices of the incomparable Paul Dourish and the LUCI Lab crew. I am thankful to Paul as well as to Irina Shklovski, Marisa Cohn, Lilly Irani, Silvia Lindtner, Melissa Mazmanian, and Gary and Judy Olson for making that experience such a rich and exciting one, for taking the time to think and talk about this project, and for their intellectual generosity. A second ethnographic project at NASA's Jet Propulsion Laboratory enabled me to follow up on the Rover team while working at the lab and to check my assumptions about how spacecraft teams work, for which I thank Robert Pappalardo.

The Society of Fellows and the Sociology Department at Princeton University gave me the precious time and space to devote to writing, revising, and editing this manuscript. Scholarly engagement with Paul DiMaggio, Mitch Duneier, Michael Gordin, Mary Harper, Erika Milam, Susan Stewart, and Paul Willis inspired fresh thinking about the project. I am thankful for the friendship and close readings of Michael Barany, Michaela DeSoucey, Amin Ghaziani, Simon Grote, Christina Halperin, David Reinecke, Molly Steenson, and Sarah Thébaud. Outside Princeton, colleagues Lisa Messeri, David Ribes, Lucy Suchman, Jennifer Tucker, and Fred Turner have been thoughtful interlocutors throughout the process. The generosity of the Barr Ferree Foundation Publication Fund and the Princeton University Committee on Research in the Humanities and Social Sciences allowed this book to be printed with color illustrations.

I owe many thanks to Karen Darling at the University of Chicago Press for her support of the project from first sight in 2007. Several of the friends and colleagues listed above read and offered insightful feedback on an earlier draft, and three anonymous reviewers gave detailed, thoughtful suggestions for strengthening the manuscript. Selections from the book appear in other venues, where they have also benefited from peer review: elements from chapter 3 appear in *Representation in Scientific Practice: Revisited* (MIT Press, 2014); from chapter 6 in *Social Studies of Science* (Sage Publications, 2012); and from chapter 7 in *Documenting the World* (University of Chicago Press, forthcoming).

I want to express my sincere thanks to all the Mars Rover mission members who contributed to this project. I owe an enormous debt to Steve Squyres and to Jim Bell, whose welcome, interest, generosity, and patience were instrumental for my fieldwork and writing over the years. Several team members endured ongoing conversations beyond meetings and interviews, especially Emily

Dean, Scott Maxwell, and Eldar Noe, as well as James Ashley, Diane Bollen, Paul Geissler, Jeff Johnson, Kim Lichtenberg, Jeff Moore, Steve Ruff, Michael Sims, Rob Sullivan, and Dale Theiling. I thank them all for their commitment to and enthusiasm for my project and for sharing their many special occasions with me, from weddings to robot funerals. Thanks also to Ray Arvidson (Washington University, St. Louis), Phil Christensen (Arizona State University), Bill Farrand (Space Science Institute), Ken Herkenhoff and Jeff Johnson (US Geological Survey), Jeff Moore (NASA Ames Research Center), Ron Li (Ohio State University), Alistair Kusak (Honeybee Robotics), John Callas (JPL), and Mark Powell (JPL) for welcoming me to their institutions as a visitor and enabling me to tour their labs, observe how they conducted their science, and interview so many of their students, staff, and faculty colleagues. I also thank Cindy Alarcon-Rivera and Mary Mulvanerton for their invaluable assistance with managing access restrictions. A full list of interviewees is in appendix B.

Throughout this adventure, my family's support, passion, and love have been unwavering. To Catherine, Les, Cam, and David, and to Craig Sylvester, much love always.

INTRODUCTION
Seeing Mars and Drawing Mars

On a cold April day in 2006, two robotic explorers on Mars awake to receive their commands from Earth. The twin robots, nicknamed *Spirit* and *Opportunity*, are NASA's latest emissaries to the Red Planet (fig. I.1). Equipped with spectroscopy equipment, a rock scraping tool, and nine digital cameras, these Mars Exploration Rovers were built to find geological traces of past water on the planet's surface. Although constructed to last only ninety days in the harsh Martian climate, they have so far survived over seven hundred days and will clock thousands more before their missions are done.

The rovers may be alone on the Red Planet, but they are commanded at a distance of millions of miles by a team of scientists and engineers on Earth, who together make decisions about where the robots should drive next and what they should do. This particular April is a critical one, since *Spirit*'s landing site in Gusev Crater, a few degrees south of the Martian equator, makes the robot particularly susceptible to the changing seasons. As the Martian winter approaches, the sun's position in the sky lowers to the north. The engineers must park the rover for the season, somewhere where its solar panels will face the dwindling sunlight and collect as much precious energy as

Figure I.1. Mars Exploration Rover. Courtesy of NASA/JPL/Caltech.

possible to fuel its heaters throughout the winter, keeping its electronics warm and protected from damage by the cold.

On this same April day on Earth, then, the Rover Planners, a team of specialist engineers at the Jet Propulsion Laboratory (JPL) in Pasadena, California, are poring over hundreds of images of the region that they commanded *Spirit* to take. They are looking for a "winter haven" for the robot—a rise in the terrain nearby where the slope will keep the rover's solar panels naturally tilted toward the winter sun as it tracks across the Martian sky. On finding a location and naming it McCool Hill after an astronaut recently lost on board the space shuttle *Columbia*, the engineers and scientists on the team agree to drive *Spirit* to that area.

But on its way there, *Spirit*'s wheels dig deep into a reddish brown patch of sandy soil and grind to a halt. The rover is trapped. The clock is ticking: if *Spirit* cannot make it to its winter haven in time, it will not survive the season. There

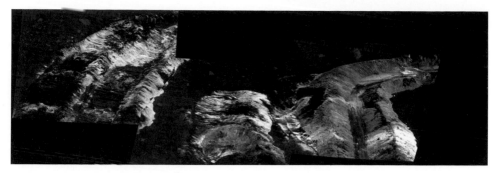

Figure I.2. Tyrone, Pancam filter 2 (753 nm), *Spirit* sol 788. Courtesy of NASA/JPL/Cornell.

is an additional complication: about seven hundred days into what was expected to be a ninety-day mission, the rover's right front wheel jammed at an awkward angle, never to turn again. The engineers must now drive the robot backward, and gingerly at that, dragging its stuck wheel.

The engineering team struggles to free the crippled rover, driving back and forth over the Martian terrain. As they do so, the scientists order the robot to take pictures of the sand beneath its wheels so they can analyze the soil to find a way for the rover to get out. As the days pass, it becomes clear that *Spirit* will never make it to McCool Hill in time, and the team members scramble to find an alternative winter haven. When they finally extricate the vehicle from the sand trap, before driving to a small ridge a few meters away and parking for the season they command it to take one last picture (fig. I.2) of its roughed-up tracks etched in the Martian soil with its stereo, full-color Panoramic Cameras. The crisis has been averted for now.

While *Spirit* sits still for a few months and *Opportunity* is driving several kilometers toward Victoria Crater on the other side of the planet, members of the Mars Exploration Rover (MER) team on Earth shift their focus to related projects. At JPL the Rover Planners convene in their "test bed," a site designed to simulate Mars, to practice with an Earthbound rover how best to drive *Spirit* with only five working wheels. At an Ivy League university on the other side of the country, the lead scientist for the rovers' Panoramic Camera instrument puts the finishing touches on a spectacular coffee-table book of Martian images; the Principal Investigator balances teaching his popular freshman course with visits to NASA Headquarters in Washington, DC; and both punctuate this work with frequent speaking engagements around the United States and Europe. Participating Scientists at private and public universities, at research centers like the Smithsonian, at NASA centers, or at the US Geological Survey head out to places like the Río Tinto in Spain, the Atacama Desert, and even Antarctica to conduct research in Marslike environments. Scientists who serve on the Long Term Planning subgroup call each other to discuss orbital images of the area and agree on how best to drive *Opportunity* to its goal at Victoria Crater, or on which direction *Spirit* should drive when power levels rise again. A flurry of e-mails over the mission Listserv circulates drafts of papers, posters, and abstracts for comments and contributions before they are sent off to *Science* or *Nature* or to meetings like the yearly Lunar and Planetary Science Conference or the American Geophysical Union Conference. And five days a week this far-flung team of scientists and engineers dials into meetings on a teleconference line to check in with the rovers and with each other, to request specific observations from *Spirit* and *Opportunity*, and to plan each rover's operations over the next few days.

It was while *Spirit* was parked for this Martian winter that Susan, one of the mission's scientists based at a private university in the midwestern United States, decided to learn to work with the rover's full-color Panoramic Cameras: the Pancams. A physicist by training who builds spectrometers to study the chemistry of soils, Susan was attracted to the chance to complement her work using two of the rover's spectrometers—the Miniature Thermal Emissions Spectrometer (MiniTES) and the APXS Alpha Particle X-ray Spectrometer—with the Pancam's imaging capabilities. She traveled to the Pancam headquarters to spend time with the operators there, to train for a role of Pancam Downlink Lead (reporting daily on the status of the remote instrument), and to learn to use the Pancam image-processing tools. During her training she practiced her newfound skills on the pictures of the patch of roughed-up soil, now named Tyrone after a county in Ireland. Shortly afterward, Susan suggested at the daily teleconferenced planning meetings that the team reconsider Tyrone as one of the top priorities for investigation once the winter was over and solar power was up. The rest of the team had little interest in returning to what they saw as a dangerous sand trap and were instead discussing moving west to explore the nearby plateaulike region they had named Home Plate.

As an ethnographer working with the Mars Rover mission, I was sitting in on the teleconferenced science meeting in October 2006 when Susan made her first presentation about Tyrone. It was not a particularly momentous occasion: all members of the science team, whether professors or graduate students, staff scientists at universities, or civil servants at NASA centers, are regularly encouraged to share their work in progress with the rest of the team at these weekly meetings before the findings are published. I sat alone in the darkened room in the astronomy building at Cornell University, a room with carpeted walls and no windows but outfitted with a Polycom device for videoconferencing. Surrounded by darkened computer workstations, since it was late in the workday, I listened to the voices on the line and watched the slides scroll by on the large projector screen on the wall, aware of other team members following along from their offices down the hallway. Susan was the last on the agenda for the day, after a long discussion of results from the rover's spectrometers. Her thirty-two PowerPoint slides, displayed over the team's live-streamed videoconference screen and circulated by its document-sharing site, started with two Pancam images that *Spirit* took of Tyrone while the rover was trying to escape the sand. The images quickly flashed from black and white into vivid false color, painting Mars in pinks, yellows, and greens.

Using these and other visual transformations of the same images, Susan argued that while *Spirit* was struggling to escape from Tyrone, its stuck wheel had exposed some light-toned soil that was different from the rest of the reddish brown soil in the area. Further, her colorful images demonstrated that there were two kinds of white soil, that they were some kind of salt, that one possibly was deeper than the other; and that the soil turfed up from the deeper layer was changing over time to share spectral characteristics with the soil from the upper layer. The presentation took over an hour, and at the end one of her colleagues laughed as he called it "the visual equivalent of drinking from a fire hose." But the group members acknowledged that they could see the two-toned soil she pointed to and found it intriguing, and they discussed taking further images of Tyrone from their winter haven position.

A few months later, in February 2007, I joined the Rover mission's Participating Scientists as they came together for a face-to-face meeting at the California Institute of Technology in Pasadena. The agenda was packed with presentations of ongoing work by science team members, their graduate students, and assistants. Questions flew from the audience at every presentation. Susan's talk was moved to the last day of the meeting to make time for a discussion about *Opportunity*'s upcoming exploration of Victoria Crater, but when she finally took the floor, the audience was riveted. In her presentation, Susan took three ways of showing the two-toned soil and applied them to eight pictures of rover tracks from across the region; she then mapped the location of these tracks to make a claim about the light-toned soil's stratigraphic location and possible provenance—as a waterborne salt deposit.

Suddenly the team members not only saw the two-toned light soil, they saw it everywhere. They were so excited by the presentation that the Principal Investigator extended the agenda for an hour-long discussion of the light-toned soil. Scientists around the room rapidly traded hypotheses about what the soil was, where it came from, and what observations would be required to resolve those questions. Is it a salty deposit laid down by water? Is it layers of volcanic deposits from a recently active volcano? When exposed to the atmosphere, does it change chemically and turn red to look like the top layer of Martian soil? Suddenly this was no longer just Susan's observation: this was the "Light Soil Campaign," a series of observations to investigate the light soil's provenance and dispersion in the region, and it was one of the mission's highest science priorities.

After the meeting, NASA issued a press release including a color picture of Tyrone and announcing the discovery. Despite the danger of getting stuck in the sand again and the pressure to move westward to the nearby region called Home Plate, the science team requested that, as soon as *Spirit* had enough power to move, the Rover Planners immediately drive it back to Tyrone for more observations. Over the course of the coming year, I watched as the team used the rover to investigate the region and compile enough evidence to claim that the Home Plate area had once been a hot spring, not unlike those at Yellowstone National Park: a discovery of past water on Mars. This earned publication in *Science* magazine as one of the most significant discoveries of the mission. The images that the *Spirit* rover returned from the Tyrone region were critical not only to deciding where and how to drive the robot, but also to conducting pioneering scientific research on Mars.

Working on the Mars Exploration Rover mission is a highly visual experience. Visual work suffuses the team members' interactions with the robots and with each other. Large full-color panoramic photographs decorate their office walls; bright false color images circulate among science team members, embedded in PowerPoint files; black-and-white photographs of Mars are painted with colored swatches to show where and how the robots might drive or conduct observations; and students spend hours calibrating raw image data files so they can be used for scientific investigation. Without images of Tyrone or of any other part of Mars, it would be impossible for these scientists and engineers to claim to discover anything at all on the Red Planet.

Yet the digital images that return from the surface of Mars do not depict the planet as human eyes would see it. Instead, the rovers' purpose-built cameras have specially selected filters so they can photograph wavelengths that human eyes cannot necessarily detect or isolate. The scientists and engineers on the mission use these filtered images that the rovers return from Mars to constantly compose and recompose different visions of the Martian surface. The mission is so suffused with this kind of work that a mission member once explained to me that joining the team required learning and developing a special kind of visual expertise: with the images the rover returned to Earth, the software suites required to manipulate them, and the common visual transformations that circulate among the team. As she described it, "When you work with the team for a while, you kind of learn to see like a Rover."

This is a book about what it means to *see like a Rover*: that is, how scientists and engineers on Earth work with the digital images their robots take on Mars to make sense of the distant planet and work together to explore its surface. Based on over two years of immersive ethnography with the Mars Exploration Rover team, I will reveal the planning, interpretation, and circulation of digital images on the mission. I will follow scientists at their desks as they perform the active manipulation and composition of digital images that make sense of a distant planet and make it available for robotic interaction. I will describe how their colleagues, too, come to see features of interest and use their digital resources at home and robotic teammates millions of miles away to develop scientific facts about the Martian surface. Throughout, I will explore and explain how the iterative and contingent activities of drawing, seeing, and interacting with Mars produce the unfolding narrative of robotic space exploration. At the same time as work with digital images of Mars produces new ways of seeing and interacting with the planet, I argue, seeing like a Rover binds these scientists, engineers, and robots into a single collective team.

Scientific Images in Social Context

Our understanding of Mars has always been subject to our imaginations. From maps famously picturing canals on its surface to Orson Welles's *War of the Worlds* broadcast, it seems that everyone sees what he wants to see on the surface of the Red Planet. In the late nineteenth century, competing mappers such as Nathaniel Green, Percival Lowell, and Giovanni Schiaparelli applied terrestrial cartographic methods to Mars, squinting through their telescopes to faithfully depict the canals they saw on the planet.[1] Introducing the photographic camera to these sizable telescopes did not so much disprove the existence of canals as expose the ambiguities of the planet's surface.[2] It was not until the Mariner missions in the 1960s first flew past the planet and took photographs with vidicon cameras that earthlings began to see a terrain unlike the one in their imaginations, and even more varied.

In the hundreds of years before the rovers arrived on the planet's surface, then, theories about Mars, practices of observing, and techniques of scientific imaging came together to produce visual knowledge about the planet in ways that historical figures considered rigorous and scientific. The same holds true today. Although our tools are robots and digital cameras, we confront similar questions. What is the role of human observers, with their observations and experience, in crafting scientific knowledge about another planet? What role can or should instruments, software scripts, and computers play in crafting this knowledge, and when should human sensibilities and experience intervene to "check" the machines? And how can we trust what our images tell us, especially when they are subject to manipulation and interpretation?[3]

This book examines these questions in the context of a twenty-first-century mission to Mars. But while my case study is Mars exploration, what is at stake is our understanding of images in science more generally. It is all too easy to assume that scientific images show exactly "the things themselves as they appear"[4] without paying attention to the considerable work it takes for scientists to produce such pictures. In this book, then, I will shift analytical attention from the images themselves to *the work of scientific representation*. How do we make objects scientifically visible, and to whom? Which characteristics of an object are included and which are excluded?[5] And how does the image reflect the values of the community that made it? Precisely which aspects of an imaged object are revealed and which are hidden, and why and how, is crucial to understanding the role of images in scientific practice, on Earth and on Mars.

Scientific seeing is not a question of learning to see without bias. Instead, scholars of scientific observation remind us, it entails acquiring a particular visual skill that allows a scientist to see some features as relevant for analysis and others as unimportant. As philosopher of science Norwood Russell Hanson put it, when Kepler and Ptolemy look to the east at dawn, they do not see the same thing. Although they both observe the sunrise, Ptolemy would say he sees the sun moving around Earth, while Kepler (a Copernican astronomer) would say he sees Earth moving around the sun. In such moments, Hanson reminds us, "there is more to seeing than meets the eyeball."[6] It takes a particular kind of training to learn to see like Kepler, like a scientist—or like a rover.

This training involves learning some degree of context: background assumptions that dictate which aspects of the scene are relevant and how these aspects are related to each other. Anthropologist Charles Goodwin calls this *professional vision*: learned techniques of observation, specific to different professions, through which we make meaning out of what we see. Whether one is an archaeologist learning the exact colors and textures of soil samples or a lawyer interpreting a video recording in court, professionalism includes learning how to recognize particular details and how to distinguish relevant information.[7] Further, we do not see with our eyes alone. Surgeons, for example, use their hands and eyes in concert, along with scopes and other visual assistive technologies, to perform complex operations.[8] Learning to see requires both bodily skills and instrumental techniques.

If scientific seeing is skilled seeing, then scientific *imaging* is skilled work as well. Anthropologists and sociologists who studied scientific laboratories in the 1970s and 1980s noted that scientists rarely see their objects of analysis without some kind of optical instrument, inscription process, or visual representation. These analysts therefore paid considerable attention to microscopes, protein gels, neutrino traces, field guides, and graphs.[9] Their counterparts today must also contend with screens, software, image files, and a range of digital visual technologies.[10] Indeed, on the Mars Rover mission, the work of scientific observation is tightly linked to both digital imaging and practices of visual interaction. Without images, Rover scientists would not have any visual experience of Mars. Without digital image manipulation, they would not come to see the compositional or morphological details of the Martian terrain that interest them. And without distributing their image manipulations among the team, they would not produce the shared visions of the Martian terrain essential to deciding where the rover should go and what it should do next.

The work of digital image processing is important not because it can produce a more perfect vision of an object under investigation. Instead, scientists use digital images to perform a wide variety of transformations, with each mouse click revealing new aspects of the object that were invisible before. They conduct a kind of work with visual materials *such that* we can see: a practical process of visual construal. They resolve potential ambiguities by focusing on one set of salient features, relationships, or objects. They build context and aspect into an image, discriminate foreground from background and object from artifact, such that other scientists come to see the object of interest the same way. They use image manipulation to convey this visual experience to the image's observers. The visualizations that result are designed, as sociologists of science Karin Knorr-Cetina and Klaus Amman put it, to "carry their message within themselves."[11]

I call this practical image craft *drawing as*, a turn of phrase that focuses on how scientists and engineers compose and recompose the same images of Mars into a variety of visual forms. The resulting images are not in competition with each other, but rather reveal and conceal different aspects of the planet for different purposes. I use the word *drawing* intentionally, here. Although I will be describing twenty-first-century work with digital images, similar practices of visual construal are present in other times, places, and media of scientific visualization, as I will discuss. It is the work of *drawing as*, I argue, that carefully constructs particular embodied and instrumental visions of the surface of Mars, brings scientists and engineers together in the process of exploration, and ultimately enables team members together to see like a rover.

Why and What Do Rovers See?

To describe what seeing like a Rover entails, it is important first to describe the robots, their provenance, and their capabilities. The rovers did not appear on Mars out of the blue, after all: their design and their implementation were shaped by historical circumstance and by individuals on the team. This history determines not only what the rovers can see, but also who regularly sees through their "eyes."

Where the Rovers Come From

The history of the Mars Exploration Rover mission—indeed of NASA's contemporary Mars program—began in 1996.[12] This was the year that a NASA scientist announced his discovery of what appeared to be traces of biological materials in a Martian meteorite, recovered in Antarctica twelve years earlier. Although NASA's Viking missions to Mars in the 1970s had not discovered any biological markers, closing the case for follow-up missions, the meteorite discovery galvanized the scientific community, their colleagues at NASA headquarters, and even the president of the United States, prompting a push for a return to the Red Planet. In 1996 President Bill Clinton announced a special program for Mars exploration, setting aside a funding stream to send a series of spacecraft to Mars. The first few missions NASA flew under this banner were constructed during the "faster, better, cheaper" era of mission management: an agency-wide attempt to curb costs.[13] The missions failed spectacularly, and NASA instigated a reorientation of its Mars program to avoid such public embarrassments. Under the leadership of Scott Hubbard, a planetary scientist placed in charge of the restructuring, the program received a new mission statement to guide all future mission development: "Follow the water."[14]

The scientists who planned, the engineers who built, and the bureaucrats who approved the twin Mars Exploration Rovers were therefore working under a shared set of assumptions and constraints that shaped the robots' bodies and capabilities. In line with the agency's directive, the Principal Investigator and his team proposed a robot equipped with the tools of a geologist to find traces of long-gone liquid on the surface of Mars. When NASA officials selected the proposal for a Mars Rover mission from among its competitors, the agency chose to build two rovers instead of one, in case of problems with landing. NASA did not want an embarrassing repeat of the crashes of the 1990s.

Geology was a compelling choice for NASA officials, but it may seem a strange choice to those who typically associate NASA with astronauts or astronomy. Geologists have the skills to analyze rocks and planetary terrain so as to make claims about the prehistory of an environment, including its past water conditions. And geologists have played a central role in NASA's space exploration initiatives since the Apollo missions. At that time, US Geological Survey geologist Eugene Shoemaker turned his disappointment at not using scientists as astronauts into developing a rigorous program to train the selected Apollo crew members in how best to find samples of lunar rock and return them to Earth.[15] His contemporaries also leveraged geologists' skills at interpreting aerial photography of Earth-based natural resource sites into skill at deciphering orbital photographs of planetary surfaces. During this early era of space exploration, then, geology formed one of the core sciences in the new interdisciplinary field of planetary science, which later incorporated atmospheric sciences, biology, and astrophysics. The design of the Mars Exploration Rovers therefore drew on a long lineage of the centrality of geology and geologists in planetary exploration.

For these reasons the two robots, each about the size of a golf cart, were outfitted with a suite of scientific instruments to approximate a geologist's tool kit. These instruments include several spectrometers to analyze mineralogical composition through spectral signatures; a Rock Abrasion Tool to grind away the weathered crusts of rocks so as to better view their interior composition; and no fewer than nine cameras. Four of these cameras, perched over the rovers' wheels, detect hazards in the terrain ahead (Hazcams); two take positional images to help with driving and operations (Navcams); one is a microscope camera on the robots' extensible arm (the Microscopic Imager, or MI); and two high-resolution cameras are equipped with special filters giving multispectral color capabilities. These last two cameras, the Panoramic Cameras (Pancams), produce the glorious and famous images of the Mars surface that grace magazine covers and newspaper pages. When I wrote this, *Spirit* and *Opportunity* had returned well over half a million images between them.[16]

The rovers do not make their own decisions about when and where to use these instruments, drive, or conduct observations on Mars. Although they are equipped with basic artificial intelligence (AI) capabilities to analyze Hazcam images while driving in order to avoid crashing into obstacles or driving off promontories, they are not autonomous vehicles. The members of the Mars Exploration Rover team on Earth together make the decisions about where the robots will go and what they will do, then they send these commands to the robots. Since images are centrally enrolled in this social process, this book will examine their decision making in detail. It is useful, however, to describe the team before witnessing its members in action.

The Mars Exploration Rover Team

At the time of my study, the Mars Exploration Rover team comprised approximately 150 individuals. Distributed across the United States as well as sites in Denmark, Germany, and Canada, the team includes scientists and engineers, each with different responsibilities, disciplinary backgrounds, and skills. Some are professors, others are professionals. Graduate students, professors, and postdoctoral or staff scientists use virtual tools to work alongside civil servants, robotics or software engineers, and hardware developers in private companies. The mission is demanding in terms of their time and their resources, bringing them together many times a week for teleconferences and several times a year for face-to-face presentations.

The scientists on the mission are members of the interdisciplinary field of planetary science. Bringing together geologists, chemists, physicists, astronomers, and biologists, planetary scientists attempt to understand distant worlds by combining tools, techniques, and research questions from these constituent disciplines. A few of the Mars Rover mission's Participating Scientists are staff members at NASA facilities, but most are employed by universities, public research centers, or private organizations. They all receive grants from NASA that support their participation on the mission, enabling them to achieve scientific goals, train graduate students, and contribute to the mission's ongoing operations.

Although NASA plays a central role in long-term mission planning and management, a mission like the Mars Exploration Rovers is not the product of a single agency. When an opportunity for a launch becomes available, NASA releases an announcement of opportunity (AO) to scientists at large, calling for proposed missions that fit the planned guidelines. Once the proposals are received and reviewed by a panel of experts assembled from among the planetary science community, NASA then provides the financial, managerial, and engineering support to underwrite the selected project.[17] This includes the support for associated scientists to build and operate instruments and conduct data analysis, and the contract for the Jet Propulsion Laboratory's engineers to implement their experiments by building, launching, and operating the spacecraft. In assembling his proposal for consideration, the Principal Investigator of the mission selected instruments that would fulfill the mission's stated goals. When NASA selected the rovers, then, the agency also opted to fund a team of scientists associated with those instruments (Co-Investigators). The agency later opened a call to further participation from other members of the planetary science community once the mission had been running for over a year (Participating Scientists). During my fieldwork, about half the scientists on the mission had participated since the outset, some of them even soldering their own instruments onto the robots before they were launched. The others joined a few hundred days into the mission and were fully integrated members of the team. I witnessed no significant distinctions between late additions and those selected at the outset.

Scientists are not the only participants on the Rover mission. The team also includes a community of engineers, responsible for the physical "base" of the rover itself: the circuits and wheels, the navigation and hazard-avoidance cameras essential for robotic driving, and communications functions, all supplied by NASA. Most of the engineers on the mission are employees of the NASA contractor center, the Jet Propulsion Laboratory (JPL). They are referred to as the operations side of the mission (as opposed to the science side) and are responsible for commanding the robots daily and ensuring their continued safety on Mars.[18] Like the scientists, few of the engineers who work with the rovers as operations staff were responsible for their design, construction, or launch. Although many of them have backgrounds in aeronautical engineering or spacecraft systems, operations is seen as a distinct type of practical expertise. As staff members at JPL, the engineers rotate onto the project from other missions but may be reassigned elsewhere as staffing requires. During my fieldwork, a core group of engineering team members remained with the mission, with only a few additions and departures.

During the first ninety days on Mars, called the primary mission, many of these scientists and engineers were co-located at JPL. There they lived on Mars time, a 24.7-hour day, the same schedule as their rovers.[19] After this initial period, however, the primary mission ended. NASA extended the contracts allowing the mission's operators and scientific investigators to continue their ongoing work from their home institutions in a style they called remote operations. This meant establishing video- and teleconference lines, file-sharing services, and e-mail Listservs for meetings and interactions that were once held face-to-face. The meetings I observed united the engineers at JPL via teleconference with scientists at Cornell University, Washington University in St. Louis, Missouri, the Smithsonian Institution in Washington, DC, and NASA Ames Research Center in Mountain View, California, among many other sites. The mission has been operating in this mode for a considerable majority of its ongoing exploration (over ten years at the publication of this book). Rather than seamlessly transporting techniques and tools from in-person interactions to online ones, the team developed a wide variety of tools and homegrown methods for working together at a distance.

Whether during "remote operations" or primary mission, working with robots on another planet presents challenges. Scientists are so far removed from their field site that they can neither experience it directly nor calibrate their instruments to manage local working conditions. With over seven light-minutes' delay between Earth and Mars, using a real-time joystick to send drive commands to a robot is unrealistic. With the Martian day a little more than half an hour longer than an Earth day, the two planets are not always in sync.[20] And in addition to the interplanetary distance from their robots, scientists and engineers are so far from each other that they may experience difficulties in communicating effectively with their colleagues. Understanding where each robot is and making collective decisions about robotic activities is no straightforward task.

Images provide an essential way of solving these problems. The Pancams and other robotic cameras provide pictures of objects around the robots that the geologists and other scientists can analyze. They provide a sense of where the rovers are and what they can do that the rovers' drivers analyze too. Still, it is not so much the images themselves as *interactions with images* that are central to work on the mission. Because the digital images that return from Mars present multiple possibilities for interpretation and robotic interaction, it is only through constant interaction—with image-processing software suites, with teammates, and with their robots—that team members can conduct their science, operate their robots, and produce knowledge about the Red Planet. It is this visual interaction that enables them to see like a Rover.

From Visualization to Social Order

Seeing like a Rover does not rely only on individual image-making practices; it is a question of teamwork. No single scientist or engineer decides where the robot should go or what it should photograph: that responsibility is shared among the team. This means the scientist is never solitary but is situated in a social milieu. We must therefore not place too much emphasis on individual scientists' screens, cameras, or eyes, since this is not the only location "where the action is."[21] Instead, images must also be understood with respect to the interactions that surround them, make sense of them, and bring the team together. As I will argue, images in interaction are central to the production of the team's social order.

Paying attention to visualizations as they are enrolled in producing social order requires knowing something about the social context in which those visualizations are produced. On the Rover mission team, this is explicitly described as a question of a unique organizational orientation.[22] Whereas most NASA missions feature a hierarchy of scientists or a bureaucratic division of labor across multiple instrument teams, the Rover team emphasizes the twin principles of flattened hierarchy and consensus operations. All the scientists belong to a single unified science team, headed by a single Principal Investigator (PI). The team demonstrates a flattened structure with few levels of management between members.[23] The rovers' suite of instruments, known as the Athena payload, is similarly integrated. All instruments are interoperable, with datasets that can be readily combined and shared. Scientists and engineers alike are also encouraged to think of the rover as a single instrument—"like a Swiss Army knife," as the PI described it to me. Any member of the mission can use any instrument, data, or rover attribute, even the rover's wheels or arm, to accomplish scientific goals.

Additionally, Rover team members constantly impressed on me the importance of *consensus* for their mission. In my observations of their work, the narrative of consensus building and related stories such as inclusion, listening to teammates, and bridge building between groups such as scientists and engineers was constantly invoked. This does not mean that communication never breaks down, that team members never clash over differences of opinion or have trouble coming to consensus. Rather, it means that when the team engages in operating the rovers, making decisions about where they should go and what they should do, or in scientific discussions about the results of their image-processing work, there is another goal in mind alongside the task at hand: achieving unity of opinion and purpose and maintaining the commitment of contributors. This complex goal is often expressed as "being happy." The work of the mission is deeply attuned to the work of attaining this state.

The team therefore betrays a collectivist orientation consistent with commons-based, participatory, or postcommunitarian systems familiar from organizational studies of technoscientific collaborations like high-energy physics or computing.[24] Individuals must consistently work together to ensure common goals and collective assent. Decisions cannot be enforced from on high but must be discussed as a group to produce what the team considers to be the right course of action. The Principal Investigator and the Project Manager frequently remind their colleagues that they too are members of the same team; their own activities are frequently geared toward producing unilateral agreement among their science and engineering teams. Thus collectivity is a primary organizational narrative that provides grounds, explanations, justification, and accounts for members' activities.[25] Members account for this collaborative practice as the means of ensuring that they conduct "the best possible science" and always do "what's best for the rover."

Consensus sounds like a positive and democratic group orientation, although studies of consensus-oriented groups have revealed the trade-offs of committing to unilateral agreement. Meetings can become endless as everyone is required to have a say; some individuals dominate conversation while others remain silent; compromise agreements that please everyone are too watered down to be effective; and minority voices are effectively silenced as the pressure to agree and not speak out against the group norm becomes coercive.[26] Rover team members have developed their own internal structure, rules, and roles for combating these inefficiencies, as I will describe. Important for this book, images and image work are a central part of this process of ensuring agreement and building consensus. I will argue that how the Rover team manages and manipulates images of Mars not only produces collective knowledge of Mars, it also produces social order among team members.[27] That is, learning to see like a rover is a social achievement.

In this way, seeing like a Rover provides a counterpoint to seeing like a state, a classic formulation proposed by anthropologist and political scientist James Scott.[28] Scott describes how centralized state authorities impose visualizations and order on landscapes and peoples, contrasting this modernist vision to local knowledge and lived experience. Like seeing like a state, seeing like a Rover also requires a mutual entanglement of ordered vision and institutional agency. The resulting images enroll multiple observers in complex social relations, but these relations are oriented toward consensus, not authoritarian control. Even though rover images are disseminated by a government authority (NASA), observing behind the scenes reveals how images are enrolled in producing a collectivist visual experience: built from the bottom up, shared across the mission team, naturalizing knowledge production on Mars, and reinforcing social orderings on Earth.

A study of a consensus-based team may appear to run counter to early literature in the sociology of science, which firmly established *controversy* as an effective way of studying the politics of science. In the heat of disagreement over results, data, or interpretation, scientists are more likely to reveal the inner workings of science and depart from the usual public picture of a communal and calm group of individuals dedicated to the scientific method. Yet just because this team operates by consensus does not mean there are no politics on the Mars Exploration Rover mission. Instead, one must come to see consensus building *as* the politics—a difficult human task that requires just as much backstage discussion, argument, subtlety, respect of interests, attention to communication, and concession as any other form of political organization or expression.

Images are one of the central products and currencies of this activity. And the interactional techniques that visually construe the planet in one way or another are enrolled in continued observations and decision making. Instead of hunting for the moments when consensus breaks down and politics rears its ugly head, then, I chose instead to treat the building and management of consensus as the social work that is enmeshed with the work of conducting science on Mars. For although the team members spend much of their time discussing the rovers, caring for them, planning for them, and managing them at an extreme distance, the social dynamic that presides over these activities ensures that managing the rovers is at the same time a question of managing the team. As Liz, a camera operator on the Rover team poignantly expressed it, "After those rovers leave Earth, the team is all we've got."[29]

Ethnographic Work on Mars

My own work with the Mars Exploration Rover team began in early 2006, when I joined one of the laboratories affiliated with the mission as an ethnographer.[30] For most of my fieldwork from 2006 to 2008 I was based at a large research university where one of the primary Rover mission labs was housed. From there I observed the daily videoconferences (SOWG meetings, described in chapter 1) at which scientists and engineers made decisions about rover activities. I also observed the weekly teleconference science meetings (End of Sol meetings) at which scientists discussed their results and engaged in what they called long-term planning for the robots.[31] To deepen my understanding of the mission, during the summer of 2007 I visited ten institutions affiliated with the mission, ranging from NASA centers to universities to private companies, to observe the planning process and the scientific work there.[32] I also accessed historical material through oral histories, publications, and archival materials available at NASA history offices and libraries. I conducted interviews with more than eighty scientists, engineers, graduate students, and staff in their home institutions and observed, recorded, and later analyzed their scientific work with digital materials at their desks. Consistent with sociological convention, names have been changed in the text to protect confidentiality; a full list of interviewees is provided in appendix B.[33]

Ethnography is not only a question of collecting a mountain of interview transcripts, video recordings, or observation data. Unlike social science methods that describe, taxonomize, or represent a community from the outside, ethnography is very much about being there. Conducting ethnographic research involves learning native ways of doing things, drawing boundaries, making decisions, and making sense of what goes on. This requires a thorough immersion in a community that cannot be obtained through periodic visits or interviews. In this case it required my full immersion as a member of the team. Like other team members, I regularly attended meetings, conferences, and celebrations, where I became a familiar face. Over time, I became as conversant as any other team member with the scientific questions of the mission, the techniques and materials used to conduct this science, and the unfolding narrative of the robots' explorations.[34] Although I did not participate in operating the robots or conducting scientific work, I did work as an image calibrator to contribute to the mission. However, team members often described my work of observing them as equally valuable, as much a contribution to the mission as the scientific findings, and they frequently referred to me as a member of their team.

I conducted my ethnography over what corresponds to approximately a full Martian year, or about two Earth years. I joined the team during *Spirit*'s third winter, as the robot was exploring the east side of the region known as Home Plate, including the discovery at Tyrone, the exploration of Mitcheltree Ridge, the analysis of the Kenosha Comets, and the negotiation over Winter Haven 3: the activities that led to the discovery that the Home Plate region had once been an ancient hot spring. At the same time, on the other side of the planet, *Opportunity* approached Victoria Crater. I observed as the team spent the Martian year commanding the robot to drive onto each promontory around the crater's rim, examining pieces of ejecta, meteorites, and wind streaks nearby. As I left the team, *Opportunity* had just begun a three-Earth-year drive to Endeavour Crater, while *Spirit* headed into the robot's third (and last) winter. I attended follow-up meetings, conferences, interviews, and Team Meetings until July 2011, when I also attended *Spirit*'s funeral. I discuss these episodes in more detail throughout the book; interested readers will find driving maps illustrating where the rovers were during these stories in appendix A.

I draw on four types of primary materials. The first includes recordings, transcripts, notes, and photographs from my attendance at over two hundred Science and Operations Working Group meetings and fifty End of Sol meetings, as well as the annual Team Meetings and other conferences. The second comprises interviews, collected at a variety of times and locations and with as wide a range of team members as possible, particularly during my site visits to mission-affiliated locations. The third consists of videos, notes, and photographs of Rover scientists working with digital visuals, analyzed with an ethnomethodological[35] approach to appreciate the routine and practical grounds of digital work. The fourth type includes field notes and experiences that record the more intangible aspects of ethnographic engagement: hundreds of informal conversations, postmeeting drinks, e-mails, and asides that simply come from *being there*. I have tried to interweave as much thick description as possible into my narrative to bring my readers into the middle of the action.

Despite my integration into the team, there is an important limitation to my fieldwork that influences this book. Foreign nationals are not permitted to access technical details about spacecraft design or operations under United States law as set forth in the International Traffic in Armaments Regulations (ITAR). Not being a United States citizen, I was limited to discussing and witnessing the scientific side of the mission, avoiding any discussion of technical details of the rovers and their operations and all situations when the latter details would be discussed or displayed.[36] As an ethnographer I take these regulations extremely seriously, since their disobedience could harm my participants. Despite these restrictions, the degree of access I was permitted generated enough material for an extremely fruitful study, and I am deeply grateful to mission coordinators under whose generous permission I was able to participate in this mission to the extent legally possible.

This necessary and embodied attention to the politics of the mission augmented my experience in the field by forcing my attention to issues that the team finds critical in the practice of planetary exploration today. Yet it also produced an artifact in my analysis in the way I deploy "science" and "operations" (or "science" and "engineering"). These terms mirror the team's own distinction between decisions, image data, and people involved in scientific work with images versus the decisions, image data, and people related to the movement or management of the spacecraft. They should be read that way in this book as well: not as an analytical distinction I draw, but as the team's own categories that sort and make sense of their activities and their membership. They also demarcate the domains I was able to explore in detail versus the aspects of the mission to which my access was limited.[37]

Being located at a distance from the detailed technical work of operating the rovers revealed a range of Earthbound practices that made sense of the robots' experiences and trajectories: the people, software, images, bodies, and screens on Earth required to craft knowledge about Mars. Included in this work are the team's own practices of making the rovers themselves present and accountable through images, embodied actions, mediating systems, and talk — what sociologists Karin Knorr-Cetina and Urs Bruegger call the work of "appresenting."[38] Ultimately, then, my story is not about the Rovers as coproducers of objects on Mars, or the exclusive role of the apparatus in crafting knowledge about the planet, but about the sociotechnical, material and organized work on Earth that constructs Mars as knowable and interactionable.

Holding our focus steady with the science team on Earth presents several advantages. It exposes the continued importance of visual practice in the context of contemporary knowledge production. These practices are material, embodied, and purposeful, aiming to make sense of the ambiguous world that is Mars and the distant object that is the rover. It also allows us to analyze the stories actors tell about their robots as part and parcel of this sense-making. To whatever extent the rovers' bodies or Martian realities shape the ways people on Earth work with them, what is evident to the Earthbound ethnographer is how people's narratives about the rover,[39] their understandings of its operation, their ways of working together, and the concomitant visual conventions they develop for viewing Mars have a powerful effect on shaping knowledge and practice. As I will show, even a Rover team member's sense of what it means to see like a Rover is bound up in many layers of visual and collective practices on Earth.

An additional advantage of this analytical position is that it reveals the role of the team's social organization both in managing the rover and in crafting knowledge about Mars. I take as a first-order assumption that the rovers cannot be understood without the complex network of people and software on Earth that animate them. What remains to be explored, however, is exactly how this team is organized, what daily work the members perform, and how their ordering on Earth is implicated in producing order on Mars.[40] The approach I adopt here reveals how the organization of individuals and their practices on Earth together contribute to the robotic exploration and scientific understanding of Mars.

Outline of the Book

All ethnographic projects face the problem of artificially organizing, for the purposes of argument or exposition, experiences and themes that are in practice intermeshed and entangled. To best reveal the role of images in interaction in scientific work, I have chosen to follow the images through the mission, from backstage to the foreground: from their planning and acquisition, to their calibrating and cleaning, to the manipulation that reveals new aspects of the terrain, to their central role in determining the robots' course of action, their constraints, and their release to the public. Each chapter presents an ethnographic snapshot of the process with implications for interaction, decision making, and scientific knowledge. Throughout, I will maintain a dual emphasis on how images in interaction represent both the planet Mars as well as the Rover team. That is, interaction with rover images does the work of ordering Mars, configuring it for robotic intervention and scientific knowledge. But at the same time, the work of ordering the field is also, crucially, the work of ordering the team as well.

In chapter 1 I go backstage to the first step of image acquisition: the daily planning meetings. Because all activities on Mars must be carefully budgeted in terms of bits, time, and power constraints, attention to these accounting practices and their negotiation among team members demonstrates how Martian imaging is a practical activity that must be carefully managed and accounted for. Because of this accounting, each image is acquired and tailored for a specific purpose, presenting implications for image processing and visualization. But images also arise from and feed back into a micropolitical process of achieving consensus among team members. In describing the work of image acquisition, then, I describe the local organizational modes, practices, and narratives of the collectivist Rover team.

Once an image has been acquired, it must be cleaned before it can be considered trustworthy and prepared for analysis. In chapter 2 I draw on material from my participant observation experience with the Pancam calibration crew to examine how and why requested images must be calibrated before team members can begin to work with them as scientific artifacts. As images are altered to approximate standardized conditions on Mars, we will see how the management of both human and machine interventions generates a sense of digital trust and objectivity.

In chapter 3 I move to the scientists' desks and screens to witness how they conduct scientific investigations with rover image data. Visions of Mars are the result of purposeful image construal: the work of crafting images so as to see a particular or even a novel aspect of the imaged object. In this chapter I show how visual transformations effected with digital tools foreground different aspects of image data such as slope, atmospheric opacity, soil composition, and rock morphology. As we follow scientists working with image-processing tools to tease out an aspect of the imaged object for further access or intervention, I will develop the analytical framework of *drawing as* in close detail. That is, I will show how the Rover scientists use digital tools to *draw* Mars *as* consisting of different kinds of materials or surfaces, with implications for future viewings and for team relations.

In chapter 4 I show how annotated images are an integral part of the same *drawing as* activities. That is, annotating images of Mars for either scientific or engineering analysis produces maps of Mars construed for particular kinds of interaction. Such maps imprint categories and kinds onto the Martian landscape but are ultimately aimed at producing further robotic interactions with Mars. In chapter 5 I describe how this image annotation figures into the group's collective discussion of data interpretation and rover activities. I describe how the interactional norms of visual interpretation produce and reproduce the local organizational orientation toward collectivity, even as they also require different team members to offer their own discipline-specific views of Mars.

In chapter 6 I turn to the embodied gesture, narrative, visual conventions, and forms of talk that accompany visual skills and techniques for seeing Mars like a rover. Team members do not so much use their robots as extensions of human senses; rather, they acquire an embodied sensitivity to the robots' capabilities, mediated through Earthbound visual transformations. Stories that circulate on the mission and shared gestures create a close, even totemic relationship between human team members and the rovers, binding team members to their robots and, through them, to each other. Through attention to these accounts and activities, I argue that the bodily practices of visual sense-making and other associated interactions play a central role in reproducing the team's collective orientation.

In chapter 7 I examine how working with Mars as an object of analysis requires disciplining Mars Rover scientists as the subjects of that analytical work too. Images cannot be *drawn as* just anything at all: they are subject to local norms and "constraints" that both delimit visual transformation and circumscribe scientific claims and persons. I therefore describe the mathematical practices and analog work in the field on Earth that scientists produce and invoke to generate claims they consider trustworthy. Ultimately, community constraints on visual interpretation shape both the subjects and the objects of analysis on Mars.

In chapter 8 I move away from the team-centered view to place image work in the context of the public patronage networks of NASA and planetary science. I analyze a type of image for public release—usually glossed with the caption "What it would look like if you were standing on Mars"—as an appeal to continued political survival in the alien wilderness of the Red Planet. Work to produce these images is somewhat distinct from the scientific and operational concerns of the Rover team. However, through digital image-processing practices that *draw* Mars *as* the new frontier, I show how team members construct a community outside their group to ensure the public's continued support and engagement and the enduring appeal of their mission.

Where Do Images Come From?
Planning a Day on Mars

"JPL, are you on the line?"[1] The sound of ringing telephones punctuates the darkened room in the middle of the astronomy building at an Ivy League university, where a local group of Rover team members gather around a conference table. Screens are everywhere: a ring of computers lines the room, topped with official signs labeling them "Pancam PUL" or "Pancam PDL," a large hanging screen on one wall displays a projection of video conference activities,[2] and it seems everyone in the room has brought a laptop. Visible on the projector screen, their colleagues at the Jet Propulsion Laboratory file to their seats around a U-shaped table in a bright, spacious room with a model rover in the center, facing a series of screens onto which shared images are projected and distributed online. Bleeps sound on the teleconference line as team members phone in from offices, cars, or coffee shops around the world. This is the daily meeting of the Science and Operations Working Group—the SOWG[3]—at which the scientists and engineers on the team make decisions about what the rover should do the next day.

Crafting a plan for each Mars Exploration Rover mission robot is a dynamic, collective, and carefully managed art that balances a

variety of competing factors. The rovers do not conduct science or see by themselves. Each day they receive detailed instructions from their human team on
Earth about where to go and what to do. And though Rover team members joke
about getting the "keys to the rover," there is no joystick that controls real-time
operations.[4] Because it can take up to twenty minutes for a signal to reach the
planet, the team communicates with its rovers only once a day, sending one to
three days' worth of commands at a time ("uplink"), and simultaneously receiv-
ing the data from the rovers' successful activities the day before ("downlink").
As a team member explained to me, "We're working on the Martian night shift."[5]

Like all spacecraft, the rovers have only so many bytes, hours, and watts
available on board with which to take and store images and transmit them to
Earth. The SOWG meeting, usually convened once a day for each rover, is the
place where scientists and engineers must balance several competing pressures
and produce a plan for both rovers' activities on Mars the following Martian day,
called a sol. The goal of this morning meeting is to produce a plan that will be
uploaded to each rover at day's end, with a sequence of observations and drive
directions that will direct its activity on Mars. Because surface situations change
daily and new scientific questions may arise on the spot, this detailed daily planning ensures that the rover wakes up to a complete list of requested activities,
compiled and negotiated day to day, with no bytes to spare.

Despite the dynamism of managing a wheeled vehicle on Mars, the SOWG
meeting is also highly routine, ritualized, and practical. With its tightly sequenced and adhered-to combinations of reports, discussions, and statements,
the meeting is a refined interaction ritual that organizes social activity on Earth
even as it achieves the goal of producing robotic activity on Mars.[6] As I will describe, the ritual character of the SOWG meeting makes certain resources and
interactions available to team members as they work together each day, reinforcing and reproducing their organizational commitment to consensus. Referring
to both the tightly scripted procedures and the need to satisfy multiple groups
and different interests, one team member described the meeting to me as "a
finely tuned little dance that we do."

But the work of managing the rover is also the work of managing the team.
The ritual pattern of SOWG interactions enforces and reproduces the local
norms that govern participation on the Rover mission.[7] Activities, roles, and
even specific conversations are enacted daily through video and teleconference
links to carefully manage rover health and activity and produce a plan for each
robot within an hour. These ritual framings, in turn, shape team negotiations
and the eventual observations the robots perform. In particular, the plan cannot

be approved and implemented until all team members at the SOWG meeting achieve consensus. Images of Mars, enrolled and produced through these interactions, constitute and reflect this social order as well.

Into the SOWG: "A Finely Tuned Little Dance"

"Can I get a roll call on the Meet-Me line?" asks the SOWG Chair, kicking off the meeting exactly on the hour, according to the clock on his computer. Remote participants state their names on the teleconference ("Meet-Me") line as the engineers file to their seats in the room reserved for Rover operations at NASA's Jet Propulsion Laboratory in Pasadena.[8] At the base of the U, a row of desks is reserved for the Rover Planners, specialist engineers who are responsible for producing the code that commands the rovers. A Mission Manager, the engineer who maintains oversight over the rover's operations for the day, is also in the room, as well as an engineer whose sole focus is the rover's status, keeping track of its changing solar power situation or communications needs.

The SOWG participants' tightly delineated roles include specific tasks and responsibilities.[9] Around the U-shaped table at JPL are blue placards that identify the liaisons from each of the rover's Athena suite of instruments: the Pancams, the MiniTES, Mössbauer, Microscopic Imager, APXS, and RAT. These team members are responsible for the current status of their individual instruments, including whether yesterday's sequences ran to completion or sent back any data ("downlink"). They may also code the instructions for the plan that will be sent up to the rover at the end of the day ("uplink").[10] At the virtual table are the mission's Participating Scientists, with their staff or graduate students who maintain close involvement with the mission. All attendees at the meeting share online access to documents posted on a secure networked site and to a live video feed from JPL showing the SOWG room and two screens displaying PowerPoint presentations and Maestro, the rovers' in-house science activity planning software (fig. 1.1).[11]

The Rover team's consensus model of operations does not mean the SOWG is a free-for-all or an endless meeting. On the contrary, the meeting's goals, structure, and roles are explicit and closely adhered to by team members. The purpose of the meeting, as they put it, is to produce a plan for the rover's activities that balances the robot's health and operational concerns, such as conserving its power or taking care not to cause damage, with scientific requests for images and other observations. Team members talk about this in terms of managing "rover health" alongside "squeezing out every last possible bit of science" from the robots. Because the SOWG also requires consensus by the end of the hour, this

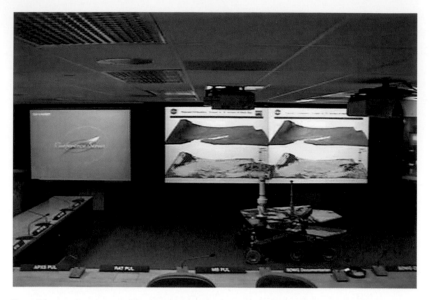

Figure 1.1. A webcam view of the SOWG conference room, the "Callas Palace." Author's photo.

means balancing competing science or engineering needs, maintaining strong working relationships between team members, and keeping the peace in case of disagreement.

The meeting is presided over by the SOWG Chair, a position that rotates among a select few scientists on the team.[12] The Chair is responsible for moving the team from an analysis of the downlinked data, indicating where the rover is and how it is doing, to a plan contingent on that analysis that will be uplinked to that rover at the end of the day for execution over the next sol. One SOWG Chair, James, explained his job to me as balancing the details of the daily plan with long-term expectations, "trying to make sure that you can dovetail the engineering and the science requirements" and "trying to get a rhythm where you cover some distance, stop and do scientific analysis and observations, then cover some more distance."[13] The tensions he points to lie at the heart of the planning. Rover team members balance this tension by articulating two distinct planning categories: "strategic" and "tactical" operations. Strategic planning is producing long-term visions of what the rover should do over several days or months at a particular location. I will discuss this process in chapters 4 and 5. Tactical planning, on the other hand, is the job of the SOWG: determining precisely, to the minute, what the rover should do on Mars tomorrow.

Much like rituals that distinguish the sacred and the profane, the ritual of the SOWG meeting reinforces a strict distinction between the strategic and the tactical. No negotiation can take place in a SOWG meeting that actors judge more strategic than tactical in nature. For example, during a SOWG meeting where the day's plan concerned an approach to a nearby wind streak blowing off Victoria Crater, a scientist spoke up on the line to register his concern with driving into the crater based on the wind streak direction. While the topic—investigating a wind streak near the crater—was related to the rover's present position, other team members were uncomfortable with conflating those issues that would become rover interactions tomorrow (wind streak investigations) and those that fed into longer-term investigation (going into the crater). The SOWG Chair immediately intervened with the request to "reserve that discussion for today's End of Sol," the meeting where strategic plans are negotiated. The ritual distinction between the two was in full force.[14]

Meetings ritually begin with a roll call to hear who is participating that day. But as roles rotate among team members, this also serves as a *role* call as participants state which operational role they are responsible for that day. Scientists clustered into Science Theme Groups called Atmospheres, Geochemistry, or Geomorphology each designate a member to be present at the SOWG meeting to represent their group's interests, concerns, and requests. Each instrument also has a Payload Uplink Lead and a Payload Downlink Lead (PUL and PDL) who closely follow the conversation and ask questions along the way to make sure group members understand what activities are requested of them and can question the requester before they spend the rest of the day writing commands for that operation. In addition to these specific liaisons, team members can also assume roles designed to encourage holistic thinking about the operation of the rover and the team in concert. The Keeper of the Plan (KOP) is in charge of entering observations into the rover's software, called Maestro, in sequence and as decided by the entire team. The Documentarian keeps a careful record of each observation, who requested it and why, and ensures that all commands and requests issued at the beginning of the day are accounted for by the end of the day.[15] An engineer at JPL is responsible for considering how all the commands sequenced by different operators will interact, so that no observation or move will contradict or conflict with another. And a group of scientists are designated as Long Term Planners (LTPs), whose job it is to stay attuned to the strategic plan, keeping the bigger picture in mind while the daily meeting focuses on the immediate concerns of rover operation.[16]

With so many different roles and responsibilities and so many potentially conflicting needs, the team has developed a particular practice for ensuring that all members feel included. They call this "listening." William, a SOWG Chair, stressed the importance of this practice to me with the local aphorism, "It could be that person is right only 10 percent of the time, but if it's that 10 percent, then you'd better be listening." The Chair engages listening at specific points in the meeting by directly inviting responses from team members, opening the floor to comments. For example, at the outset of a divisive discussion about where the rover should go, William initiated the conversation with, "I wanna hear everyone express their view."[17] Members recognized this prompt and used it as the opportunity to speak up and declare other points of view, voice alternatives, and question assumptions.

Listening plays several important roles on the mission. It demonstrates to team members that whoever is in charge is merely a moderator and may not make a decision or close a topic of conversation without the assent of others. This reinforces the ideal of the flattened hierarchy and participatory engagement. It was also explained to me as a way of ensuring unilateral agreement through participation. After a SOWG meeting that I observed at his side, James explained:

> At the end of the meeting you want to people to have a sense of ownership of the plan. That's why I kept asking at the meeting, "Are there any other comments, are there any other comments?" . . . It's the whole empowerment thing, the team needs to feel like they're part of the process, and they're getting their two cents in and we're doing the right thing. . . . That's the most important thing, because if you wait to the end [of the meeting] and everyone comes in with their own discipline-oriented or pet peeve kind of things, then it's chaos, total chaos.[18]

Inviting comments is "the most important thing" because it reinforces the idea that James, as Chair, is not calling the shots. It demonstrates his interest in eliciting minority perspectives to be heard, if not always acted on. Further, the result is a team that "feels like they're part of the process" and therefore will approve the resulting plan at the end of the day. Note that the "empowerment" afforded by structured moments of listening is juxtaposed to "chaos, total chaos," a breakdown in social order. James described this chaos in a further contrast:

> There are kind of two ways to do the plan. One is that you send atmospheres guys, the geochemists, the geologists, the whatevers off separately to come up with their druthers, and then you make sure it fits. It won't fit. And there will

be hard feelings. The other is [the Rover way] that you start off with a strategic [long-term] plan that people have bought into, and then you give them realistic constraints as a group, and then help them develop a tactical plan that fits into the strategic plan. . . . And where people's observations can't be fit in, you develop a liens list [a "to do" list] and make sure they understand that we're gonna get to them.[19]

James's description highlights the importance of continued buy-in to the plan among team members, enacted through practices of listening that allow them to "get their two cents in" and "make sure they understand" that those requests have been heard. It is through these collectively oriented work practices that James ensures not only that there are no "hard feelings," but also that the SOWG meeting will remain orderly, without descending into "total chaos."

Formulating Place

Following roll call, routine presentations update everyone in attendance—whether in the room at JPL, in a car in Flagstaff, or at the kitchen table in Ithaca—on the rover's location, health, and immediate challenges. An initial, very brief statement from the SOWG Chair sets the stage for the sol:

We can see the rock target—and again, correct me if I've got any of this wrong—but it looks like we're at Cape Faraday, a small rock shown here. It is reachable and it is RATable [amenable to the Rock Abrasion Tool], and just to remind folks, the importance of making this measurement . . . is that we got a very unusual chemical composition last time we imaged at a trench, and we want to find out if we're seeing . . . a correlation between this rock and the . . . high magnesium sulfate composition.[20]

In another example, the Chair reminds the team of the previous day's failed observations so as to establish today's "tactical situation":

We had an IDD [Instrument Deployment Device, the robotic arm] fault when we went to [use the Microscopic Imager]. . . . It looks like we got a bunch of MIs [microscopic images] that were not anywhere near the target and are still out of focus. . . . So my plan for today is to actually recover the MI [images] . . . then bump back and look at [the target] with Pancam. . . . That's a summary of the current tactical situation this morning.[21]

Statements such as "it looks like we're at Cape Faraday" do not imply impersonal geographical points on a map but refer to a local nomenclature shared among members of the team.[22] Frequent use of "we" (instead of "us" versus "them") identifies all the participants on the line as members of the unified mission team, engaged in a collective process. Whether team members are physically in different rooms or have just finished a shift working with the other rover, the forms of talk in this report orient the team within the rover's frame of reference to establish a shared position with the robot on Mars and shared membership in the planning event. As such, they serve as opening statements that formulate a sense both of place on Mars and of participation in a conversation on Earth.[23]

Knowing where the robot is located is the first step. Following this brief presentation, the SOWG Chair invites a Long Term Planning (LTP) representative to frame today's "tactical" considerations in the context of the team's longer-term "strategic" objectives in the Martian landscape.[24] This LTP report is an image-heavy PowerPoint presentation. Updated daily with recent Navcam, Pancam, or Hazcam data, the images remind the team where the rover is, which science targets are of importance in the scene, and which overarching goals must drive the formulation of the plan. For example, in a typical SOWG meeting, *Spirit*'s LTP Lead put up a slide showing the east side of Home Plate (the area the rover was exploring) annotated with arrows and labels to indicate where the rover was, what it had already accomplished, and what it had left to do:

> The second slide will remind you what we have been doing, give you some context. . . . [We are located] between those two green arrows which define the existing Pancam coverage, so we've already imaged from position labeled "first," and we hope we are at the position labeled "second," which we hope will enable us to finish off that gap, and we expect to turn around and take images of Mitcheltree Ridge and then finish off the observations of Mitcheltree Ridge before driving up the onramp. . . . We're about four meters from the outcrop that we wanted to image, and so the idea was to bump forward maybe two or three meters so we can get better images and MiniTES observations.[25]

Similarly, on the approach to Victoria Crater, *Opportunity*'s LTP Lead[26] kicked off his presentation with a panoramic view of the crater freshly downlinked from the rover's Navigation Cameras and a view from orbital imagery to give the team a sense of "where we currently are" and "where we're heading" (fig. 1.2): "This is the map view image from where we currently are, and we are at that open green dot, and our target is that light green dot at Duck Bay. . . . I

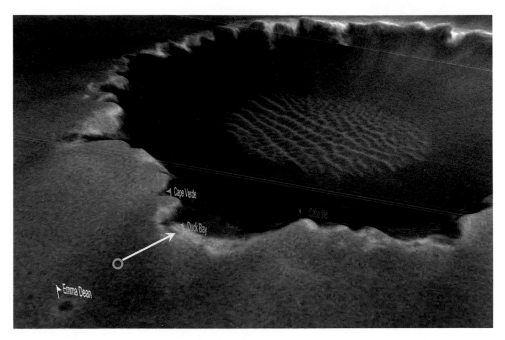

Figure 1.2. "Where we are and where we're heading." *Opportunity* LTP Report, September 22, 2006. Used with permission.

put the light green dots to give you a sense of where our target is, where we're heading on Monday."

The LTP report concludes with a slide describing solar energy fluctuations[27] and the status of the rover's flash memory for that day.[28] Based on this information, the LTP Lead offers a recommendation to the SOWG Chair: a limit of how many bits can be in the day's plan. This might be anything from 30 to 350 megabits, and it changes daily depending on factors that range from local dust storms to the location of the communicating satellite as it passes overhead. This number is important: SOWG Chairs must attempt to fit the day's requested observations into the recommended bit count. This may involve merciless cutting or smooth finessing in the sixty minutes to come.

Full constraints for observation planning are announced at the end of the LTP report, after brief reports are requested from instrument representatives and from the Rover Planners to build a picture of the rover's "health": its status, current situation, and any recent malfunctions.[29] An engineer then presents "the skeleton": an Excel spreadsheet outlining precise times when the rover must "sleep" or "nap"

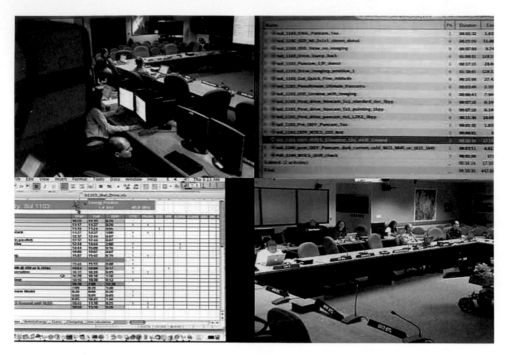

Figure 1.3. Video feed showing Maestro activity planning software (top right) to keep track of the timing and duration of activities as they are negotiated, and "skeleton" (bottom left) to show what times and durations are available for observation planning. Author's photo.

to recharge its batteries, when it must communicate with Earth based on satellite passes overhead, and what time is available to the science team to request observations. This outline presents a frame into which all planned rover activities must fit or be adjusted (fig. 1.3).[30] As a team member explained to me, "The skeleton is the engineering constraints that give us the time and energy within which we have to do our science." Just as no strategic discussion should take place during a SOWG meeting, no observation can be planned that violates the amount of power available or timing for communication periods. These are the parameters within which any scientific observations—"science" for short—must be planned.

The introductory sequence of the SOWG meeting articulates roles on the mission and the structure for interaction in which these roles are embedded. Presenting the skeleton, making instrument reports, describing solar power fluctuations, and reviewing both context and recent history not only ensure that all team members are on the same page as they move into the negotiation process, but also provide resources for managing the open negotiation of specific observations that will ensue: from vehicle health, location, available bits, and time available for scientific observations to which participants are on the line today and the requests or concerns they will bring to the table.[31] These images and conversation about them at the outset of the meeting further establish a distinction and a back and forth between the strategic and the tactical. As the LTP report and the skeletons leave the screen, the Keeper of the Plan loads their view of Maestro, the science activity planning software (also called SAP), to the remote display for all to see, and the Chair opens the floor to requests for observations.

"Mind the Bit Bucket": Observational Accounting

The top consideration when requesting observations of Mars in the SOWG meeting is, will it fit? What an observation must "fit" into is that changing target of recommended bits, watt-hours, and timing based on changing conditions on the Martian surface, satellite availability, and the rovers' own flash memories. To meet this daily "bit quota," especially if there is a drive involved or solar power is low, the team members together account for every bit of data requested from the rover. No observation can be made willy-nilly, and no time should be left over with nothing for the rover to do. Team members describe this as "Minding the bit bucket."

The language of rover planning is thick with accounting metaphors.[32] Chairs frequently speak of "tallying" bits and "bookkeeping" observations. Another explained the process to me as "a system of checks and balances"[33] by which the numbers of bits, watt-hours, and observations were traded. For example, looking over the list of activities planned for the day, a Chair noted an observation with the navigation cameras that would require more time and bits than the rover could afford that day: "One thing that we still have on the books is this Navcam. . . . Right now I'm still bookkeeping the standard twenty-five minutes; we're gonna have to think about what we want to do with the data products there."[34]

Such vocabulary—bookkeeping or being on the books—echoes across the SOWG. In this typical example from the outset of a meeting, the KOP summarizes the observations inputted to the planning software (Pancam, Mössbauer, Hazcams, and driving) with reference always to "bookkeeping" time or data space for those activities: "Sol 1100. Engineering Pancam Tau, and then I have bookkept the IDD Mössbauer touch. . . . I have a two hour and forty-five minute drive . . . then we have the ultimate and penultimate [drive] Hazcams . . . we have the . . . post-drive Pancam, 4 by 1 three bits per pixel . . . and that's it "[35]

The actor's metaphor of "accounting" evokes the sense of a finite amount of resources, and of trade-offs that must take place within a tightly delineated space. The "bit bucket" of rover memory can accommodate only so many observational requests before spilling over. So requesting images usually involves what team members call an "advocate": someone who can speak to exactly what the observation is, what it requires, and why it is needed. This advocacy may be based on scientific considerations, but it must also conform to the language of accounting to explain exactly why each bit, second, and watt is necessary. Advocates may simply speak up to make the request during the meeting once the Chair opens the floor. Or they may e-mail the KOP before the meeting requesting an observation of a particular target of interest; even so, they will still be required to justify their request during the meeting. The following scientist's request for an observation is a representative example: "I'd like to advocate for one of these quick L2 R2 pan[orama]s. . . . If there's a hole in the [sol] 957 plan we should put it in there. It's on the order of fifteen minutes and something like under twenty megabits."[36]

Note how the scientist articulates exactly what he wants and when he wants it: a panoramic image with the second filter on the left Pancam (L2) and the second filter on the right (R2), only fifteen minutes long and twenty megabits of memory, which he believes can be easily placed into an existing temporal gap in an upcoming sol 957 plan. Presenting an observation with a description of how many bits and minutes it will require, or if the time of day is crucial, is part of the language of accounting that argues for incorporating an observation into the plan.

After a particularly long satellite pass or during a sunny Martian day, flash memory can be fairly empty, and SOWG Chairs may simply open the floor to the scientists to suggest observations. For example, one Thanksgiving an atmospheric scientist was the only one on the line owing to his colleagues' holiday travel. The SOWG Chair therefore jokingly proclaimed the day "Christmas for Atmospheres" as the scientist made request after request. On capping off the day with a bit-heavy "cloud movie" imaging sequence, it was renamed "Atmospheres Gone Wild!"[37]

Usually, however, memory space on the rover is tight, so all requests are subject to detailed scrutiny by the Chair to make sure that bits in the "bit bucket"[38] are "bookkept." This results in the strategic editing of observations to fit the amount of time, bytes, and watts the rover has available for the day. The observation's advocate is then required to stick up for a request or identify just what can be cut should there be too many bits in the eventual plan. In such cases, if the advocate has not already clarified the purpose of the observation, the SOWG Chair will ask exactly what the image is for, to tailor the observation specifically to that need alone and free up time, watts, and bytes for others. Team members call this process "trimming."

For example, faced with a request for a Pancam observation using the left and right cameras, a SOWG Chair tried to articulate exactly what the observation was for: "So lemme ask about these Pancams. What you got right now is a Pancam L7 R1 two by one [frame mosaic]. . . . I think what we want them most for is . . . identifying with confidence where there is exposed rock."[39]

Identifying what the image is for permits the SOWG Chair and image advocate to trim the image to just its predicted context of use in order to conserve bits. "Trimming" the image resolution (measured in bits per pixel) requires walking the line between staying under the bit limit and producing an image that is still legible for its purpose. In this case the camera operator and the scientist negotiated the trade-off between how many bits they could spare for the day's plan and how high the resolution had to be to fulfill the goals of the observation:

> Camera PUL: To get the best focus, going with three [bits per pixel] is the best bet, if we can afford it.
>
> Scientist: If you do need to drop back to one [bit per pixel], the 20 millimeter is the best position.[40]

Resolution is one image attribute that can be trimmed; the choice of filters is another. The Panoramic Cameras have thirteen filters, but it is costly in terms of spacecraft time and resources to regularly use them all. More frequently, scientists must decide exactly how many they need for their observations and often have to be satisfied with no more. For example, while *Spirit* was driving toward Home Plate, Cynthia, a geologist, requested a thirteen-filter observation of a strange feature in a rock along the way. Confronted with more bits in the proposed plan than proposed in the LTP report, the Chair held the observation up to scrutiny:

> Chair: Cynthia, that's great that you have this observation in there to look at this [ridge in the rock]; it's potentially a really interesting target, but I guess I'm just wondering what's the rationale for thirteen-filter rather than L257R1 LOCO [four photographs: the left Pancam's second, fifth, and seventh filters, and the right Pancam's first filter, all low compression] . . . if you're trying to characterize the dust . . . ?

Cynthia: No, I guess I'm curious to see if it's something different [than we've seen elsewhere]. So, well, I guess it doesn't have to be thirteen-filter.

Chair: Would you be happy also with an L257R1? Just if we are tight on bits.

Cynthia: Yes, I guess that would be fine.

Here the Chair trims the proposed observation from using all thirteen Pancam filters to only using four (the second, fifth, and seventh on the left Pancam eye and the first on the right) with low image compression. This, he believes, will be appropriate for Cynthia's purpose, given that she is interested not in seeing the dust, but rather in determining a difference in composition. As I will discuss in chapter 3, trimming the observation to accommodate specifically what Cynthia is interested in seeing results in images through certain filters that can be combined to reveal some geological distinctions but not others. It thus marks the beginning of the *drawing as* processes that will shape later representations, observations, and interactions.

This kind of trimming occurs regularly across the mission. The prevailing philosophy is that planning the right images requires knowing what purpose they will be put to. Team members openly acknowledge this limit and deploy it as a resource in proposing their specific images. In the example above, a scientist who was in fact interested in seeing the dust spoke up to question whether limiting the observation to the filter set L257R1 would be appropriate given what she also wanted to see:

Alexa [to Chair]: This is Alexa, I'm just wondering what is L257R1 going to reveal in terms of differences. What would you predict? . . . Cynthia wants to find out if this is different spectrally from the surrounding, and I'm trying to get at if it's dustier. What are we going to be seeing? What's different?

Cynthia: I want to get a sense of whether it's the same composition and maybe go visit this thing at some point.

Chair: I think an excellent example of what L257R1 can reveal is [the targets] Montalvo and Riquelme. We imaged those with L257R1, and you can see very clearly the color and texture differences. . . . We have a long track record of using these filters to distinguish between different units.

Alexa: Okay, then.

Chair: And I want to emphasize the textural.

Alexa: I'm all for texture.[41]

The Chair's assurance that L257R1 is appropriate is not just pro forma, intended to placate Alexa and Cynthia. L257R1 is considered a particularly useful filter set across the mission both because it is constrained enough to plan regularly without taking up too many bits and because it enables scientists to accomplish a range of image-processing transformations. Having a right- and a left-eye image will allow them to compose a stereo picture to emphasize texture, while the L257 filters cross a wide enough range of the visible light spectrum to detect many compositional differences.

Although the observation has been cut, the interaction represents what the team would consider a successful negotiation. The Chair has clarified that the limited filter set can still accomplish its advocate's scientific goals, the observation has been trimmed so that the science fits into the day's recommendation, Alexa and Cynthia have voiced their concerns, and a potential misunderstanding between the Chair and the two scientists has been verbalized and resolved. As for the resulting image to be taken on Mars, transmitted to Earth and analyzed, it is suited to a particular purpose yet ambiguous enough to permit a modest number of transformations. Thus such images are planned bit by bit through interaction and negotiation among members of the team.

Resolving Tensions

These examples required scientific rationales: to see the chemistry and the texture of a ridge or to identify where there is exposed rock. But not all images are "for science"; many are "for operations," a catchall phrase that includes driving the rover, placing the robotic arm somewhere, or monitoring solar power. For example, driving the rover requires an entire suite of images: a navigation camera mosaic of the rover's prospective drive direction, images taken along the way to ensure that it is proceeding in the right direction, and images taken at the final location so that the Rover Planners know where their vehicle is.[42] Should a drive be included in the day's plan, the Rover Planners request images to support that drive. These images can be "bit heavy" and take up much memory, but they are considered "mission critical." They are given high priority for acquisition and downlink to Earth so they will be available to the SOWG by the time of the next meeting, usually the next day, for purposes of immediate planning. They may even bump scientific observations off the activity plan, as in the following example: "We got some pretty challenging driving ahead of us and if we've got only two sols to get those images down, then we can go with more images, lower compression ratios, more bits per pixel, just give the Rover Planners better-quality products so they can do what we're gonna ask them to do on Monday."[43]

In another case, a SOWG Chair realized late in the meeting that while they had planned many scientific observations, they had not included the drive images:

> *Chair:* Given that we've got to do a short bump [i.e., drive] . . . do we already have the drive direction Pancams and Navcams that we need to plan that drive?
> *Rover Planner:* Nope.
> *Chair:* I was afraid you'd say that.
> *Rover Planner:* We can possibly get by without the Navcams, but we can't get by without the Pancams, and we probably should get the Navcams.[44]

In this case the scientific activities planned for that sol were cut to make room for the Pancam and Navcam images necessary for the drive. However, operations images are more frequently managed alongside the science requests and are therefore not always set in stone. When the terrain appears predictable or if the Rover Planners are set to drive the rover a long distance, the Rover Planners frequently offer to take lower-quality images to conserve extra bits for scientific imaging. But when the terrain is more complex, the imaging they require can be intensive enough to require removing scientific observations from the plan. When a Chair observed that the recommended bit count for the day was used up before they could even get to science requests but still optimistically began including observations, a team member explained to me off-microphone, "The science will just get cut [later]. . . . Too bad, isn't it?"

The trade-off between scientific and driving observations throws into relief a relationship that Rover team members believe requires constant management: that between scientists and engineers. Stories abound on the team about fractured relations between the two camps on previous missions, and about the importance of paying attention to methods of communication between the two sides. Scientists, as characterized in these stories, always request too much without an understanding of what the spacecraft can and cannot do; engineers, for their part, are described as too protective of the vehicle and, as the saying goes, "would prefer to fly a spacecraft with no [scientific] instruments on it!"[45] The limited number of bytes and watt-hours in a plan can serve as a resource to manage relations between the two groups. If the rover can permit only a certain number of bytes in a plan, and the plan requires drive images to get to where the scientist wants an observation, then the Chair can appeal to the needs of the rover to pacify the advocates of the competing observation. Similarly, the scientists' or engineers' generosity in offering ways of trimming their own observations builds rapport between the two sides of the team, goodwill that is often repaid with similar generosity in later meetings. Thus policing the bit quota is a resource for managing both proposed observations and the team members who proposed them.

Rover health and safety were frequently invoked as a reason engineering images needed to take priority over scientific images. At the time of my field-work, the rovers were already running hundreds of sols over their recommended ninety-day limit, but the team still treated the threat of rover death as imminent. Not taking an image that could protect the rover from injury constituted an unacceptable threat. But death was invoked as an argument for scientific obser-vations as well. For example, as a scientist suggested trimming to a three-filter instead of a thirteen-filter Pancam observation, the SOWG Chair refused, say-ing, "I don't know what the science objective is for it . . . but I'd hate to see us drive away from this spot and whoever it was who wanted this doesn't [get it]."[46] Instead of trimming the observation, the Chair kept it in the plan as it was, as-suming a scientific rationale for the request. In another case a mission scientist urged the team not to go "throwing away a drive sol" on a panorama of Victoria Crater in the interest of getting to the rim faster and seeing more. When his colleague protested that "driving away from this pan[orama] would be nuts," he explained, "I like imaging probably more than—just as much as—anybody, believe me; I'm just worried that we're going to run out of sols at the end."[47] The Chair reassured him that the cost of days it would require to take the panorama was "more than counterbalanced by the quality of the scene in front of us" and continued with the observation.

Such examples reveal how the threat of mission's end can be invoked as a resource to make decisions, to support or justify a proposed plan of action, or to assuage team members' concerns about their observation. They also show how SOWG Chairs are required to make decisions about imaging under the pressure of doing it right the first time, with no chance to go back and possibly no op-portunity to go forward. After another such decision cut short an observation in favor of driving onward, a team member turned to me and explained, "This is Mars: we're only here once, you know."[48]

This attitude betrays another tension on the mission: between the pressure to stay in one location and conduct detailed scientific analysis and the pres-sure to move on and see new things. Thus some images are taken and justi-fied in the plan as supporting the spirit of exploration that infuses the mission, alongside the consideration for careful science and cautious operations. The value is placed not on routine or planned activities, but on making discoveries by noticing and pursuing the unexpected, arousing and satisfying curiosity, and appreciating the sublime nature of the field setting. These images are described as "just sort of a postcard," or getting "a really spectacular image." Periodic images for aesthetic or sublime interest satisfy team members' sense of exploration. For example, the following exchange between the SOWG Chair, the Pancam PUL, two scientists, and the Mission Manager took place on a sol when *Opportunity*'s power was high and memory banks were low:

> *SOWG Chair:* Something I've been wanting to do for a long time at Victoria
>   Crater is to take some Pancam imaging . . . with the sun low in the sky.
>   There's really—there might be some science that would pop out of this,
>   but think of all the pictures that you've seen of the Grand Canyon an
>   hour after sunrise or after sunset with all those long shadows from those
>   promontories. I'm curious to see what the crater looks like at that time
>   of day . . . maybe take a few minutes to get a really spectacular image. Just
>   sort of a postcard.
> *Roger (scientist):* Sounds pretty.
> *Sam (scientist):* Could become the [NASA] "Image of the Week."
> *Chair:* This is . . . something that we're doing sort of for fun, who knows
>   maybe something good will come out of this but I'd like to try this if no-
>   body objects. . . . Suppose we did a 2 by 1 red-green-blue, everything one
>   bit [per pixel], how long does that take at sixteen megabits?
> *Thomas (Pancam PUL):* [consults computer] About five minutes.
> *Chair:* How much duration can I have for this, guys?
> *Mission Manager:* However much you need.
> *Chair:* All right, so I want five hours, now tell me what you really want.
> *Mission Manager:* [laughs] I mean, within reason. I'd put an upper cap of—if
>   it's over fifteen minutes.
> *Chair:* Let's bookkeep this for now . . . as a Pancam 4 x 1 one bit per pixel
>   L257. And if we have to change that we will [downgrade it]. . . . And
>   Thomas, you and I will just look at where the shadows are supposed to fall
>   and find the prettiest place to do it. . . . Thanks, everybody.[49]

Such moments produce a competing value to that of tight accounting for bits and a different kind of observation than the careful, trimmed activities. This is not just field science:[50] it is described as exploration. The idea of driving in a direction that "depends on what we see!"[51] and looking out for "targets of opportunity"[52] can inspire the team to lay bit counts aside in favor of opportunities for exploration or for a periodic "glory pan." This is sometimes called looking out for "dinosaur bones." Of course no bones have been discovered on Mars, but the terminology suggests "a mythical discovery that will force the science team to stop in the middle of the drive."[53] For example, referring to the discovery of a particular geological structure, an LTP Lead declared, "That's sort of our dinosaur bone. . . . This is the kind of stuff we need to go after."[54] The term argues for an observation's importance by appealing to the team's sense of exploration and serendipity. SOWG Chairs must therefore balance members' accounts that move between the language of tough accounting and that of the "spectacular" and serendipitous.[55]

A final and crucial example of these images is extremely bit-costly high-resolution Pancam panoramas, usually produced with thirteen filters, of the entire 360 degrees or a large image frame around the rover. Acquiring these panoramas can take several days or even weeks of planning and requires careful management of power and flash memory. For example, in the case above, the scene at Victoria Crater was considered to be well worth the cost of mission days to capture. Two reasons for this are typically given. On the one hand, the image can be released to the public to inspire continued interest in the mission, as I will describe in chapter 8. On the other hand, these panoramas capture something of the time and dedication it takes to get the rover to a particular location. The landmarks such images register are not necessarily places on Mars but may be significant moments among the team. Liz once explained to me that these panoramas function "to mark a significant location, to record an incredible view . . . but each one of them also tells a story."[56] Team members frequently cite panoramas like *Spirit*'s 360-degree view at Husband Hill (fig. 1.4) or *Opportunity*'s view of Victoria Crater from Duck Bay (fig. 1.6) as their favorite images of the mission, not for scientific reasons but because they capture the group's triumph on finally getting there.

Achieving Consensus

The most important ritual of planning is performed at the end of the SOWG meeting. After the negotiations for observational time and resources, the Chair calls on representatives on the teleconference line by role and asks if they are "happy" with the plan. The appropriate "response pair" to the question is a chorus of "yes, I'm happy." For example:

> *Chair:* Are the atmosphere, geology and mineralogy people happy with those observations that have been added in?[57]

Figure 1.4. *Spirit*'s 360-degree panorama view from the top of Husband Hill, recording the end of a long and difficult climb. Courtesy of NASA/JPL/Cornell.

Or

Pancam PUL: Pancam is happy with the plan.[58]

Individuals may also be called on directly to express their "happiness" with observations they requested or negotiated:

Chair: Nick sounds happy, yes?[59]

Participants recognize the ritual closing as an opportunity for them to request any information they need to do their job that day. For example:

Chair: Rover Planners, are you happy?
RP: Are we happy? Do we have a sequence number for the drive? And then we're happy.[60]

The ritual response pair has become so common that it is often subject to joking. Team members declare themselves to be "the happiest,"[61] "ecstatic,"[62] or "so happy we can't stand it!"[63] One particular role on the mission, the TAPSIE, frequently responds with "TAPSIE's hapsie!"[64] Once all members have stated their "happiness," this indicates that the planning process is over and the next stage of assembling the agreed-on commands can begin. Another way of putting this is that planning cannot proceed for the rover unless all the participants declare themselves "happy," so one of the goals of the SOWG meeting is to arrive at that moment of agreement. The SOWG Chair will not initiate the closing ritual until it is clear that all on the line will declare their "happiness." If anyone on the line admits discomfort, the group will not continue to produce commands for the rover.

On the one hand, the language of "happiness" stands in for something like "I'm satisfied" or "I have no further questions." On the other hand, as an interactional sequence, the repeated assertion of each individual's "happiness" produces a particular social effect. It requires team members to express their continued commitment to the mission and reminds them that in doing so they are all complicit in the day's activity plan.[65] This is the moment where each team member asserts not only support of the plan, but membership and solidarity in the mission as well. It serves as a reminder that the plan is ultimately subject to the approval of everyone on the line, and that success in proposing or advocating for an observation relies on keeping everyone on the line "happy." Being "happy" is, for mission members, a direct and affective expression of the team's organizational orientation and the success of the SOWG meeting.[66]

Images play an important role in this process, particularly when team members are at loggerheads. When facing a controversy about what the rover should do, a common strategy for preserving consensus is for the SOWG Chair to declare that there is too little information in the images to make an informed decision. The next step of rover progress, therefore, changes from whether to choose one option or another to a question of what further images are required to make the decision or where the rover should drive in the interim to get more visual data on which plan to adopt. For example, on the final approach to Victoria Crater there were many vivid discussions in both strategic and tactical meetings about which way to drive around the rim. The decision eventually imported to the SOWG, however, was described this way by the LTP Lead: "On Friday we start the drive toward Victoria Crater rim. We might not be close enough to see far enough to make any decisions, but the next drive, which we'll plan on Monday . . . [will bring us to] a good place to have a look at Victoria Crater and decide which way to drive from there."[67]

At a Team Meeting in February 2007, too, the team rounded off a discussion of where to go into Victoria Crater and how to get on top of Home Plate with a similar deliverable: drive to a new point and make more observations from there until the path becomes clear or evolves further. An LTP Lead even presented an annotated image that proposed moving only as far as a projected reconnaissance point, saying, "As soon as we get up there, we take an image so we know what we're gonna do [when we arrive] at the top."[68] Team members therefore move quickly from challenging an individual's position to asking each other, "Are there any observations we could make that would nail that down?"[69] Like agreeing to cut an observation to a limited set of filters or placing a target, the decision to put off a decision until there is more visual information is also considered a successful moment in the mission, another point where consensus is achieved.

It may be that the team does not have enough information to make an informed decision at the time, particularly with respect to the trafficability of the terrain. A shared idea that images relay information leads easily to the conclusion that further imaging will simply generate more of this information on which to base an increasingly informed and unified opinion. However, such information cannot be gleaned unless and until the images are disambiguated through collective interpretation. Putting the decision off until tomorrow, then, is not so much a conflict-avoidance strategy as a way to buy team members time to come to an agreement outside the limitations of the meeting format, through e-mails and side conversations. By the time the group reconvenes, tensions are usually soothed and a decision may be achieved. Indeed, while from far away which direction to go around Victoria Crater seemed ambiguous, once *Opportunity* approached the rim I was surprised to note that the drive direction decision had been resolved offline before the downlink of the new spectacular panorama from Duck Bay, the first point of access to the crater.

When Things Go Wrong

I have witnessed only one case where the SOWG meeting broke down, a breach that was notable for what it revealed about the ritual production of social order.[70] In March 2007, *Spirit* was driving away from Tyrone toward Home Plate, a feature about the size of a football field with a unique topography visible from orbit. On the way, one of the scientists on the team noticed that rocks in the region displayed unexpectedly high silica content in MiniTES spectral observations. Looking over the downlinked images from the previous week, he circulated an e-mail to some of his colleagues with an image attachment. Using PowerPoint, he circled the target areas he was interested in observing to further a hypothesis about the silica-rich rocks. This would require putting off the drive onto Home Plate for another few days, substituting a drive toward one of the relevant targets where the rover could return observations using its Pancam and MiniTES instruments.

As everyone assembled on the SOWG line on Monday morning, the LTP Lead mentioned in the group's report that over the weekend a rich discussion had taken place on the Listserv involving new potential observations. This seemed to set the stage for a discussion of how to drive toward the targets. But then a Rover Planner spoke up to talk about how to get onto Home Plate. The SOWG Chair engaged with the Rover Planner on this topic, asking for input on what he described as the "philosophical question" about the priorities for observation on top of Home Plate.

At this point, I noted in my field notes, the lines of communication in the meeting diverged. The scientists on the line started to sound antsy. Since they had decided to look at local rocks for the day and put off the drive to Home Plate, this sounded like a strategic discussion instead of a tactical one, and thus out of place in a SOWG meeting. Side conversations therefore developed, leaving several questions posed on the line unheard and unanswered. Throughout the confusion, both scientists and engineers spoke of "the drive" without articulating exactly where the rover was driving. At the end of the meeting when the KOP suggested reading out the plan before asking for everyone's assent, the SOWG Chair said, "No, I've been having a look over it myself and it looks fine." When everyone hung up, a local member worried that her colleagues were "sort of assuming that because [they] understood something, everyone understood something."

The voices on the line must have sounded different than usual, since my notes record that the Principal Investigator (PI)—although not participating in this particular meeting—came into the room several times from his office across the hall "looking worried." At the end of the meeting, when I asked what the problem was, a team member offered several possible explanations: "I think it's a bunch of people not normally working together, it's Monday morning, people seemed like they were foggy, they're doing things they wouldn't normally do, it just felt disjointed to me."[71]

This could simply stand as an example of people being "off their game," as some team members accounted for the unusual meeting after it had ended, but it had consequences for rover operations and for team "happiness." As the images came down the next day, they showed the rover perched on the edge of Home Plate, poised to ascend to the top—not poised over the rocks as the scientists intended. E-mails flew around the science team Listserv. The scientist who had identified the targets was confused about why the rover was on top of Home Plate already, but the Chair and Documentarian thought the drive had been executed as planned.[72]

As everyone dialed back in to the subsequent SOWG meeting, the mood was tense. The Rover Planners at JPL, unaware of the confusion, launched into a description of how they planned the next drive to place *Spirit* firmly on top of Home Plate—the goal that, as they understood it, the scientists had been pushing toward for months. But when a scientist started to ask how soon they could get the rover *down* from Home Plate and back to where it was the day before, the Rover Planner making the presentation faltered. "I'm sorry, I can't make out what you're saying," she said.

About fifteen minutes into the meeting, the Principal Investigator spoke up: "Can I try for a second?"

> What I'm seeing here is a little bit of a disconnect between what the science team, some of them, are saying and what the Rover Planners are focusing on. The science team has a great deal of interest, some of them at least, in some outcrops that are not on Home Plate; the Rover Planners seem very focused on how to get onto Home Plate. We want to get onto Home Plate eventually, but I think we need to listen hard to what some of the scientists wanna do before we do that and come up with an appropriate plan that achieves the necessary science before we actually get onto Home Plate.[73]

Instead of accusing one or another team member of not doing the job properly, the PI's language is vague in describing what "some" scientists and Rover Planners want to do. He also reminds the team of first principles in the performance of their ritual planning. There is a "disconnect"—the kind of thing Rover team communication should avoid. This is, further, a "disconnect" between the scientists and the engineers, a zone that the team believes is typically rife with tension and prides itself on its ability to avoid.[74] The PI articulates, as he sees it, the goals of both sides and states an order in which both of those activities could happen that would satisfy both sides. He also invokes the value of *listening* ("we need to listen hard"), which team members accept and expect as the conduit to good science. In the thirty minutes that followed, the discussion adopted an intense clarity as each side presented its concerns, goals, and assumptions, looking for points of compromise or a "location where we can achieve those multiple goals." At the end of the meeting, the Chair recapped the conversation and ended by asking if everything was understood: "So, do the Rover Planners have a better idea of where we're driving and what we're doing?" The Rover Planner replied, "I believe we have an idea of what our goal is, that is what we were looking at in the pictures of [the previous sol]." The meeting even ended with a joke: when the Chair announced the name for the new target, the scientist who proposed the observations quipped, "But 'that outcrop' was working so well!" This comment, referring to the ambiguity of language that got the team into trouble in the first place, was met with laughter all around [75]

Several issues are worth noting in this example. One is that the language of consensus and inclusion can itself be invoked as a resource when things go wrong. The PI's response to the situation was indicative: as he said, "I'm not pointing any fingers here or looking to blame anyone for anything. It's just that it's part of my job to keep everyone happy, and when something like this happens, it's helpful to me to understand why it happened." The focus thus moved to collective rallying to improve the situation, away from "pointing fingers," without disciplining the SOWG Chair or taking the reins, and maintaining an overriding goal to "keep everyone happy."

But another interesting issue arose from the meeting. After it ended, I suggested that it was unfortunate to lose a day of driving and science because of the misunderstanding. But the PI was quick to disagree. "Note the silver lining stuff!" he exclaimed, pointing excitedly to the Navcam image that *Spirit* took at the end of the troubled drive at the edge of Home Plate (fig. 1.5). The image showed fine stratigraphic layers at the edge of Home Plate stretching off into the distance, which those geologists focused on rock morphology would soon see as a tantalizing clue to Home Plate's history and formation as an ancient hot spring. The Principal Investigator continued: "This is one of the most amazing images of cross-bedding we've ever seen on this rover! [The Chair] gets it, but if Stewart [a geomorphologist] were on the line he'd be jumping up and down. This is one of the more important Pancams we're gonna take, and it's gonna be splayed across the page of some journal or scientific magazine, and it was completely fortuitous; we didn't go looking for it. . . . Exploration is like that."[76]

In the PI's account, the miscommunication was recast as a happy accident. Invoking the "dinosaur bone" language of exploration, the cross-bedding was transformed into a target of opportunity. The Pancam panorama acquired as a result was indeed eventually printed in a scientific journal as supporting evidence for Home Plate's hydrothermic origin.[77] With the use of local resources such as appeals to listening and role following, the team returned to its working "happiness" and orderly interactions.

Conclusion

Chair (PI): Hey, hey guys, when you get a chance to look at [this image] of Cape Verde, it is just stunning. It is absolutely stunning.

Figure 1.5. "Silver lining" Navcam image of cross-bedding at Home Plate. *Spirit* sol 1148. Courtesy of NASA/JPL/Caltech.

Figure 1.6. *Opportunity*'s panorama from Duck Bay, on arriving at Victoria Crater. Courtesy of NASA/JPL/Cornell.

[*Others on the line*: "Whoop!" "Oh my gosh!" "Aah!"]

Pancam PUL: I'm surprised we could get a color pan[orama] that quickly.

Chair: Oh my goodness gracious golly gumbo, this is great.

Pancam PUL: Yeah, yeah, I can't believe it, we're there.

Chair: Yeah, yeah it's just . . .

PUL: You know what I mean, I think we can declare victory.

Chair: Yeah, yeah, we made it. That was, that was a beautiful, beautiful job
[planning] yesterday by everybody, just spectacular.[78]

The room is crowded and the mood in the SOWG is jubilant. *Opportunity* has just arrived at Duck Bay on the edge of Victoria Crater after a yearlong trek through the Meridiani Planum dunes. As the SOWG Chair pulls up image after image of the closest promontories, Cape Verde and Cabo Frio, the team members on the teleconference line chatter excitedly about what they see, which features are geologically exciting, and where they might drive, take pictures, and deploy the rover's instruments next.

The usual understanding of the popular expression "a picture is worth a thousand words" is misleading in this case. Pictures do not always speak volumes for themselves, but since they must be planned, negotiated, annotated, discussed, transformed, and spoken for, we might say that thousands of words go into crafting them. Rover images in particular are the product of daily negotiations and interactions, scripted and improvised, between members of this group of distributed scientists and engineers. A remarkable image like the panorama of Victoria Crater (fig. 1.6) not only is dramatic, it is the culmination of a detailed series of plans and negotiations. Like all images on the mission, this panorama

was proposed and advocated as an observation, calculated, bookkept and accounted for, approached through a drive, and requested as a Navcam mosaic before being used to target filtered Panoramic camera observations. It was also planned in an effort to resolve a disagreement on the team about which way to drive around Victoria Crater, ensuring the continued production of consensus and the chorus of "I'm happy" at the end of the previous SOWG meeting. It took thousands of additional images to produce this single view: not only the individual frames that make up the panorama, but a digital trail of images that were displayed, annotated, dissected, planned, disputed, and ultimately agreed on.[79] Only then could they be downloaded to the "oohs" and "aahs" of the team, and displayed in color at a NASA press conference a few days later. And only then could these pictures begin to circulate among the team as the subject of further discussion and analysis in order to decide what to do next.

How rover images are constructed is bound up in their immediate, situated purpose as well as in the interactions of the Rover team. Images are both the product and the currency of this activity, and their eventual form and capabilities for image processing are shaped by these actions. The rituals and practices of image planning produce a collective orientation and the goal of continued, daily consensus. The display of current images at the outset of the SOWG meeting to get everyone on the same page, the discussion and careful collective trimming of observations, and even the selection of "glory pans" are also enrolled in reproducing this particular social order. Thus images stand both as the mechanism for achieving consensus and as the record of that same achievement. With this in mind, then, we can begin to glimpse how visualization practices not only compose and represent Mars in very particular ways, they also compose and represent the Rover team.

Calibration
Crafting Trustworthy Images of Mars

2

The SOWG meeting has ended, but the day's work is just beginning. Engineers and Payload Uplink Leads across the United States settle in to code the commands that will be sent to the rover that afternoon. I am about to start my day's work too. I follow the corridor around to the other side of the building, swipe my ID card for access, and enter a square, carpeted room with computers set up on desks in a ring around the room. Technical diagrams, plaques displaying *Science* magazine covers, and colorful Martian panoramas decorate the walls, and the center table is littered with recent issues of the *Journal of Geophysical Research*, a model of the Mars Rover, and old Mars globes. I make my way to the two Linux machines in a corner of the room, with paper printout tent cards taped to the tops of the monitors, saying MER-A *Spirit* and MER-B *Opportunity*. On logging in, I load a standard text file and start the Pancam calibration software. Instantly, images of Mars clutter my screen, mostly pictures of rocks, soil, and the sun, frequently punctuated with something that looks like a joystick protruding from a bull's-eye: the Pancam calibration target. The work I am about to do in the next four hours, I am repeatedly told, will power the scientific results of the mission.

Calibration is so central to Rover science that the Principal Investigator even saw fit to include it as a topic in his popular trade paperback about the mission: "Calibration is essential for any instrument you send into space. You're going into an unknown environment, measuring things that no one has ever encountered before. So how do you know you can trust what your instrument's telling you? . . . [W]ithout [calibration] we'd never be able to figure out what our readings on Mars meant."[1]

Whether or not calibration holds the same fascination for the public as it does for scientists, metrology is a familiar problem in social studies of science. Historians have described how voyages of discovery in the seventeenth and eighteenth centuries were plagued with questions of how to coordinate measurements and materials collected in foreign locations with the standards and legibility required by the "center of calculation" back home.[2] Sociologists who have studied contemporary instrumentation aimed at detecting controversial phenomena such as gravity waves or neutrinos have shown that no matter how rigorous, calibration fundamentally relies on the scientific community's assumptions about what constitutes evidence of the phenomenon in the first place.[3] Studies of such cases provide rich examples of how politics and social relations shape technical practices and the perceived trustworthiness of instrument reports.

Trust in instruments is precisely the issue at stake in Pancam calibration. The Pancam's centrality to the mission cannot be overstated; in upcoming chapters I will describe how planetary scientists deploy its images to conduct investigations of Mars, how images are enrolled in long-term planning, and how they inspire public engagement. Despite this centrality, calibration speaks to a particular tension about the Pancam's scientific results. Cameras are accorded a special privilege in presenting to us what appears to be an objective and factual view of the world, captured by an impassive observer. Yet those same cameras can be questionable reporters as they are subjected to alien, uncontrollable conditions. Instead of providing a direct window onto Mars through the rover's eyes, then, images are deemed untrustworthy for analysis until they have been calibrated.

Planetary exploration provides a particularly revealing case study of the role of calibration in producing trustworthy instrument reports for several reasons. Because the instrument is so far away, calibration is a question not of adjusting the instrument itself, but of adjusting its results. That is, Pancam calibration operates directly on Pancam images of Mars, retrospectively correcting them to accommodate local conditions as if they were taken by an instrument that had been calibrated in advance. So instead of producing a single stream of calibrated data that effaces the process of calibration, Pancam operators compare examples of pre- and postcalibration images. Such a comparison reveals exactly how instrumental calibration directly affects and even fundamentally alters observational results. Further, the distance between the Rover and its terrestrial teammates exposes calibration as distinct—technically and organizationally—from instrument operation and scientific analysis. The cameras are twenty light-minutes away, so no technicians can physically access the camera to tweak its parameters, apply their "magic hands" to solve a problem, or adapt it to changing experimental conditions. Calibrators are not involved in the instrument's day-to-day management, do not participate at the SOWG meetings, and do not play a role on the mission proper; their tools are similarly distinct.[4] This double isolation draws attention to the practices of calibration as a unique component of instrument operation.

With these distinct practices in view, calibration routines reveal to the analyst exactly what a scientific community of practitioners believes it takes to produce trustworthy observations. What the community approves as a proper intervention reveals much about the status awarded to instruments and to observers. Further, given that the interpretation of images relies on manipulation, calibration provides a first glimpse into how members of the team account for trustworthiness as achieved through a particular kind of digital intervention. This is especially interesting for the study of digital images in scientific practice, since calibration does not rely exclusively on humans or on machines. Instead, as I will describe, Pancam calibration relies on local, artful negotiation of human and computational interactions with image data to produce an image with which scientists may conduct scientific inquiry. This negotiation involves humans' exerting judgment over computational scripts, but it also involves training human operators themselves to be as standardized and interchangeable as code or machines—and therefore just as trustworthy. Examining the calibration process, then, reveals a particular division of labor between humans and machines—one considered trustworthy enough to discipline images of Mars and prepare them for interaction.

"We Want a Human Eye": Judgment against Computation

In the spring of 2006 I joined the Pancam Calibration Crew (PCC)[5] as a participant observer, and I continued to calibrate images for the Rover team for the next three years. To join the PCC, I underwent about twenty-five hours of training sessions followed by working with a supervisor close by in the room. PCC staff

Figure 2.1. The calibration station. Each rover is assigned one computer with two screens. Calibration procedures for both rovers usually run simultaneously. Author's photo.

explain that training is necessary because the calibration "pipeline," or routine, does not simply involve following the ten-step procedure outlined in detail on the instruction sheet; nor does it simply involve executing programs and waiting for them to run through to completion before starting the next program. It also, crucially, involves the ability to make judgments about images at each stage of the process, to determine which are acceptable and which are not, and to decide whether calibration has gone according to procedure. Such training is instructive for the sociologist too. During training sessions, otherwise tacit expectations, roles, and norms are made explicit.[6] Experts in a particular area must articulate what it takes to be expert. To that extent, my analysis in this section is informed not only by my own practice as a calibrator, but also by my own training and by observations of other training sessions.

Calibrating hundreds of images is a heavy task in terms of both computer power and visual attention. Each rover has its own dedicated Linux-operating computer, and each workstation sports two large screens, arranged contiguously so that images may spread across both displays (fig. 2.1). Calibration procedures for both rovers run simultaneously on the two machines. At any given moment in the calibration process, the human calibrator will be attuned to at least two of the four screens. Several Linux-based applications must be open at the same time. These include a standardized calibration log in a text editor; an automati-

Figure 2.2. A screenshot during the calibration process shows multiple windows of raw Pancam images open in an image viewer. Author's photo.

cally updating module indicating how many images await calibration; a terminal window to enter command-line prompts; another window to enter and execute scripts in IDL, an astronomical image-processing software; and in-house applications for viewing and interacting with images (fig. 2.2). Taken together, these systems, screens, and application windows formulate the "interactional space"[7] of the calibration station.

Visual inspection plays a substantive role in Pancam calibration. Immediately on logging into the system, calibrators start a program that shows "thumbnails" (small frames with downgraded image quality) of all the images that have returned from the rovers in the last downlink, sorted by Martian sol. Inspecting hundreds of images one at a time builds up a visual vocabulary to confirm the expected, so that it becomes easier to exert judgment by discerning problematic or interesting features in the images. In fact, this visual knowledge of Mars is one of the most important kinds of tacit knowledge transmitted during the training phase. Instructors demonstrate technical procedures, but they also talk extensively about the images they see. They try to articulate what makes an image acceptable or not, or how they know the procedures have gone correctly, so that their trainees can "get an idea of what you're used to seeing."[8]

Scrolling through images during the inspection phase, my instructor casually commented, "Typically you have an idea of what Mars looks like."[9] As a novice at the time, I found this comment intriguing: I had little idea of what Mars looked like except for the fantastic landscapes of science fiction! I soon learned that while PCC group members who had been there since the primary mission relied on their knowledge of the camera's electronics to know whether the instrument was working properly,[10] two years into surface operations, new calibrators based their judgments on a locally developed knowledge of "what Mars looks like," transmitted from instructor to student. The source of a calibrator's trustworthy exertion of judgment, then, is honed visual acumen.

This shared visual skill (a type of professional vision[11]) is essential to ensure that the calibrator will notice if something looks unusual, because, as my instructor put it, "sometimes strange stuff happens."[12] In the first stage of calibration, the instructions require students to "look through the images to get a sense of what has been downlinked" and note any "obvious anomalies." These might include visually striking changes, such as the bright glare of image saturation, the static of compression errors, or black patches caused by "data dropouts."[13] To tag these anomalies for the computer, calibrators must go through the images one by one and mark each image either "usable" or "unusable" by clicking on the thumbnail and marking any problems in a dialogue box. They must also make note of these anomalous images in their "Operator Notes" text file report, which they will send in to their supervisor at the end of the day.[14] As my instructor explained, this human monitoring was in place because "we want a human eye to look over" the images: computers could be not be trusted with the complex judgments of image quality.[15] This reveals the importance of training and experience in developing a visual acuity for images of Mars, learning the specialist kind of seeing essential to *seeing as*. But it also demonstrates an important local dividing line between humans and machines, a story that calibrators repeated again and again to account for the validity of their approach.

This back and forth between humans and machines was especially visible during the core of the procedure: analyzing the "caltargets." Short for "calibration targets," these refer to Pancam photographs of a little sundial placed at the rear of each rover (fig. 2.3). While it also has a public outreach function,[16] the sundial is specially crafted for the purpose of image calibration at a distance. The Pancam Payload Element Lead, responsible for building and operating the cameras, explained that the purpose of the caltarget was to resolve the problem of not knowing what Mars looks like. As he put it, "On Mars we cheat, we say we know what this piece is we brought with us . . . to have a bit of ground truth."[17] The dial is carefully crafted with red, green, blue, and yellow sections and three different scales of gray filling the inner circles. Each of these colors

Figure 2.3. The calibration target (caltarget) before being affixed to the rear of the rover. Courtesy of NASA/JPL/Caltech/Cornell.

and shapes was specifically selected with full knowledge of its size and wavelength before launch. Since the team knows what the sundial ought to look like under familiar conditions, this can represent an absolute value with which each image can be compared.[18] At least once a day, and sometimes more frequently, the team instructs the rover to take Pancam images of this calibration target through whichever filters are required for the day's observations (fig. 2.4). Then, by comparing the rover's daily images with known values and observing the quality of the gnomon's shadow, the calibrator can determine just how much the local conditions are affecting the collection of photons in the other images the Pancam returns.

Calibrators' interactions with the caltarget images do not allow them to determine local conditions, however. This is left to a computer. Instead, the "human eye" is necessary to compensate for the failings of computer vision. The computer can compare local values with ideal values and establish the unique metrics for each iteration of the routine. But the calibrator must identify the different colored zones on the caltarget for the computer, so that the computer can then calculate how much each individual image must be adjusted. Calibrators

Figure 2.4. A Pancam caltarget image acquired on Mars. *Opportunity* sol 641, R7. Courtesy of NASA/JPL/Cornell.

describe this practice as simply using humans to accomplish what computers are not particularly good at: visual judgment.

To identify caltarget regions for the computer, the calibrator uses a software paint tool to manually draw on each caltarget image (fig. 2.5) and tag those pixels as belonging to one or another zone of the caltarget. This hand marking of the caltarget images is painstaking and time consuming. In fact, it is so time consuming that my instructor wrote a software plug-in for the calibration tool that lets calibrators click a single button and have the computer "Automatically Select Regions of Interest." Although this might seem like replacing human labor and the very value that humans bring to the process, instead it is widely recognized (even by its author) that this program is imperfect. To be sure, a key factor in the routine is to avoid pixels on the border between two regions. If the light blue digital paint that tags the outer ring of the sundial comes too close to the inner ring, the shadow from the gnomon, or even the "Earth" painted in orbit around the gnomon, this could adversely affect the resulting routine. PCC members

Figure 2.5. Coloring in a caltarget for computer recognition. Note how colors identify regions of the target. A reminder of what the target looks like in true color is pasted on the bottom of the left screen, and a checkup graph is on the right screen. Author's photo.

are therefore highly cautious about region identification. They will run the automatic program but then spend considerable time carefully inspecting each filtered version of the same image, shaving slices off the automatically colored sections pixel by pixel so as to give adjacent regions a generously wide berth. This aspect of the calibration pipeline, itself a labor-intensive visual task, is often pointed to as one of the reasons machines can't just run the whole procedure by themselves. Human intervention and judgment are needed to overcorrect the images so that the team can always "be on the safe side." Human error is here invoked as an advantage as PCC members are trained to consistently err on the side of caution. As one instructor I witnessed told her trainee, "It's always better to get too little [of the region] than too much."[19]

Such an emphasis on human judgment versus mechanical automation recalls historian of science Peter Galison's identification of judgment as a twentieth-century virtue that replaced the nineteenth-century value of "mechanical" objectivity: the passive inscription of phenomena by recording devices.[20] According to Galison, the extraordinary range in variation in instrument reports had to be subject to an expert eye to distinguish ideal types or pathologies in individual images. But calibrators' judgments do not establish an ideal type or stand in contrast to mechanical operations. Rather, the story team members tell is one of a symbiotic relationship between humans and machines. Computers just are not very good at making these judgments of similarity and difference, but humans are, or can be trained to be. The computer can be too precise, so using humans to "get too little" shields it from making mistakes that might jeopardize the resulting calibration. Conversely, enlisting circumscribed human error on the side of caution protects the calibration routine from any damaging results of human intervention. Erring in this specific way therefore is described as part of the human calibrator's contribution to the computational routine.

"Always Refer to the Procedures!" The Mechanism of Human Work

Although the Pancam Calibration Crew is led by the scientist who designed the cameras and managed daily by a staff scientist, most of the work of calibrating is done by about a dozen students, mostly undergraduates, with an interest (not necessarily a major) in astronomy or geology. The work takes place in a lab area set aside for the senior scientist at his department in a large US research university and is supervised by postdoctoral and graduate students. Hiring undergraduates is an advantage to the program, I was typically told, because it includes young people in the excitement of the Mars mission as part of their educational experience. It also produces a workforce of invisible technicians[21] who, while they may complement their PCC work with genuine interest or technical skill, are also valued for their tabula rasa approach. They are understood to be as programmable as the machines they operate.

On joining the team, members receive fifteen pages of instructions, outlining step by step which programs to run, which passwords to input, and what to look for in the resulting images. These instructions codify and control how PCC members interact with the system and may even be seen as a highly explicit written version of a technological script that guides users' interactions with a particular system.[22] In this case the script is designed by expert users for PCC members to circumscribe their interactions with the calibrating computers and images. The instructions are evolving, and new versions come out every few weeks with additions or subtractions: students are exhorted to "Always refer to the procedures! They may change from day to day."[23] The instructions detail exactly what to do in the course of calibrating, but they do not necessarily tell students what the programs are doing or what the acronyms in use mean. This can result in some differences in interpretation of local terminology and a black-box sense of the software scripts as they are executed. However, because new members are assigned to five or six training sessions with a more senior student on the team, they frequently pick up local definitions of terms and an idea of what the programs are doing to the image files.

Like most technicians, then, PCC members develop particular local expertise and shared accounts for their activities. For example, members' accounts of the calibration target itself emphasize the certainty of visual knowledge that the calibration targets guarantee. However, these narratives also typically focus on ascertaining the colors of Mars. Explaining how and why the caltarget region painting worked, a new calibrator explained to her acolyte, "Because we know exactly, like to the wavelength, what these colors are, so we can match them." Another accounted for her work thus: "It's to help when they make the mosaics to know what the colors are."[24] While color plays a role in marking the caltarget regions, and correcting images for local conditions does produce a different sense of the color of the pictured object, the relation between identifying calibration target colors and Martian colors is somewhat indirect.

Similarly, once the calibrator has painted in all the caltarget regions, a series of graphs are automatically generated for the next phase of visual inspection. Each of the colors placed on the regions of the caltargets appears as colored crosses plotted alongside a diagonal line (fig. 2.6; also visible on the right screen in fig. 2.5). Calibrators are again called on to judge, this time how closely the different colored plots align with the diagonal line, which varies in location for each filtered image. None of the calibrators I spoke to could identify what the graph was or what it meant, or what it expressed about their caltarget image tagging, but all could tell when they had done something correctly or incorrectly. Inspecting one of her graphs, a trainee noticed that a light blue dot was a bit farther from the line than she'd have liked. She went back to the image, shaved more edges off her light blue selection identifying the outer ring of the caltarget, and returned to regenerate the graph to see if it had had any effect at all: it was unclear. And when one of my plots somehow turned up with every point on the line, my instructor for the day insisted it was "perfect" and wondered aloud how I did it.

Here again, the instruction sheet provided little indication of how the plot is generated or what the graph represents. But the invisibility to the calibrators of the code and indeed much of the calibration process created a metonymic effect whereby the proximity of colored dots to a diagonal white line had a direct, albeit seemingly random, bearing on the success of the calibration practice in general. Even those who presented a coherent story of a correlation between the colors on the caltarget and the colors on the plot could not predict how

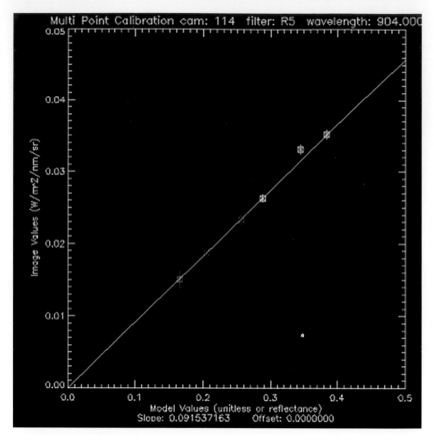

Figure 2.6. Graph with pixel values for each of the caltarget regions. Calibrators must get the plots to align as closely as possible to the white line. Author's photo.

altering the space occupied by a region on the caltarget would change the graph for the "better" or "worse." Still, a basic visual inspection, which did not require understanding the underlying processes, was enough of a check on the system to encourage trust in the eventual results.

The opacity of underlying processes persisted throughout most of the PCC members' interactions with the calibration software. While team members develop fluency in achieving their tasks, the computerized aspects of the calibration pipeline were hands-off, set in place by the team leaders and involving a bewildering array of code streaming across the screen. PCC members responded to this black-box effect in different ways. One experienced calibrator was content to stay ignorant about anything the software is doing except for the big picture.

When I asked her about a program, she answered vaguely, shrugged, and offered, "I think this is part of not knowing how [the code] works."[25] A new addition to the team, however, had already purchased a book on how to program in IDL, the language the scripts were written in. "I don't know what's going on behind the scenes," she said, "I wanna know what they do with the software."[26] This interest was fueled by curiosity and warmly welcomed by the Instrument Lead, who frequently runs an undergraduate-level seminar that teaches the coding language the machine instructions are written in and who initiated a calibration work-shop to better support the new calibrators on the mission. At the time of my fieldwork, however, only one PCC team member had taken to actually tinkering with the process. Although he is proud of his computer skills, he never hacks the calibration scripts themselves. Instead, he has written supplemental programs, such as one that automatically determines caltarget regions, that assist in the existing system.

So while calibrators may display different degrees of knowledge about the program, the technology remains a black box, even though access to its coded scripts is available for consultation.[27] Calibrators are content to let the computer run through its image transformations, limiting their interventions to those vi-sual inspections that are locally deemed to be "what humans do better than ma-chines." In general, then, human judgment is enlisted in the calibration process to check the machine but is at the same time tightly trained, circumscribed, and limited by the opacity of the algorithms. Thus the locally judged efficacy of human intervention is not due to human attributes such as creativity, ingenuity, or even error; rather, it is due to the humans' routinely operating like code too.

Confronting the tensions between the curiosity to know more and intervene in the routines versus the importance of not knowing how the code works, it is tempting to read this group as an example of a deskilled underclass labor force. Sociological studies of new technologies in work environments have described how technologies can destabilize social relations or reinforce social control over labor.[28] But the calibrators' stories and particular expertise resonate more strongly with sociological studies of laboratory technicians, such as those by Julian Orr or Park Doing. As with Orr's photocopying technicians, calibrators' local stories about calibration routines are important for group membership and for accounts of work.[29] In the synchrotron laboratory, Doing's technicians experience an agnosticism toward the experiments they operate, much as do PCC members. In the synchrotron, the "magic" was in the hands of the instru-ment operators, whose intuition for the machine was credited as a particular kind of expertise granting political status within the lab.[30] Pancam calibrators may not have magic hands, but they are trained to have particular kinds of eyes. The training grants them a relationship to the code, as human participants in a system of checks and balances that ensures results that the team trusts. So while there is certainly an epistemic politics to this particular visual labor,[31] the story of the calibrators' particular and local skills, accountings, and expertise also factors into the story of digital objectivity on the mission.

Discipline and the Digital Form of Images

The next step in the calibration pipeline is entirely digitally achieved. Calibrators type a command at the IDL prompt and are asked to enter their names before the computer takes over. Producing scrolling text on a Linux terminal screen, the actual program is invisible to the common PCC member, who is instructed, "Read a book . . . get a snack . . . this may take a while. But . . . don't go to [sic] far."[32] Although PCC members may be nursing a cup of tea, tending to their homework, or reviewing data from the other rover as this program runs, this does not mean users are passive with respect to software. All the calibration routines were produced in-house by expert programmers who continue to be senior members of the PCC and were written to accomplish the particular goals of calibration as a routine practice. This means that the repeatable scripts act on and transform the images according to a preestablished value of what a reliable image ought to look like. The raw images transform into cleaned images, producing a standardized view of Mars.

Much work on imaging in scientific practice emphasizes cleaning or otherwise imposing standardized visions on raw images. Sociologist of science Michael Lynch identifies this work as a question of "disciplining" the images: deploying techniques of mathematization or selection on elements in a busy visual field in order to display only those aspects that scientists consider salient.[33] While this has several resonances with the present case, and with the notion of *drawing as*, this story takes a twist in the software environment. Here a technology (software code) works on a digital artifact (an image file) in order to effect transformations at the level of its encoded information. The image exists only within the context of that same technologized environment, saved on the machine's hard drive as a combination of zeros and ones. The image itself, not the instrument or the artifact, changes under the force of this script. Assumptions of what makes a good image are directly encoded into the image data as the original pixels are disciplined into calibrated values.

How are the pixels digitally disciplined? Using the identification of caltarget regions, the program corrects each image according to two constants: essentially, a "lab" value and a "field" value. The "lab" determination is a radiance constant (RAD) determined during the preflight testing period "to estimate the radiometric conversion coefficients on Mars . . . to determine the camera responsivity, and assuming a 'typical' Mars radiance spectrum as output."[34] That is, it is a single standard determined before the cameras left Earth that provides some constancy in its application to all Pancam images. The second value, in contrast, is a constant generated by the comparison between the values of the identified caltarget areas and their expected values, known as IOF.[35] Thus images are adjusted primarily according to in situ calculation of what Mars is actually like on any given day, subtracting dust and other atmospheric opacity factors from the scene. The result is a duplication of each image through the calibration pipeline into images disciplined to two different calipers: one with metrics generated in the lab and constant across all images and the other generated in context in order to eradicate that context from its digital record.

Essentially, then, a calibrated image has been operated on by a software script to first identify, then subtract the effects of the atmosphere from the scene. Such a procedure is common in astronomical image processing, where indications of the location- or time-specific nature of the observations are frequently removed to produce an image that astronomers consider objective. For example, the standard routine of "flatfielding" divides images by an image of a neutral background—for example, the night sky, or the dome of the observatory in which the telescope is housed—so as to correct for any irregularities on the charge-coupled device (CCD)[36] itself. Images can then be "normalized" by multiplying them by an average value derived from the flatfield image.[37] Although it might be strange to think of Mars without dust in the atmosphere (or divided by the sky, for that matter), for the geologists on the Rover team this dust scatters the light, compromising the ability to measure the reflective properties of the surface. Dust therefore presents an aspect of Mars that must be tamed before images of the planet can be subjected to scientific analysis. RAD files, on the other hand, are often preferred by the atmospheric scientists on the team, since they preserve the measurement of how many photons actually hit the CCD in situ, and therefore the effects of dust scattering from the atmosphere. Thus, producing a clean and trustworthy image can mean drawing the dust or other changeable features in or out, depending on what the scientist wants to see. Producing a calibrated image is a question of *drawing* Mars *as* a standardized object.

Disciplining the Calibrator

Once the automated calibration routine is complete, calibrators must open the duplicated images in their image viewer once more to conduct a final visual inspection, to be sure nothing has gone wrong during the computerized process. Here the original version of the image becomes a point of reference, so that calibrators may compare the new images against their memory of the uncalibrated ones. As they do so, four more automatically produced graphs pop up, plotting pixel values in both the RAD and the IOF images and offering a visual comparison with the twenty images from the previously calibrated set. Calibrators must also visually inspect these graphs to be sure no single image is out of line with the average values in both the current and past suites of images. This can be visually taxing. For example, tau sequence images of the sun can produce sharp peaks in pixel values on the plot because of the brightness of most of the image. Calibrators must scroll through lists of file names and pictures to be sure each pixel peak on the graph indeed corresponds to a picture of the sun, or that low pixel values correspond to a night sky image.

Very rarely is it apparent that something has gone wrong with the calibration scripts, so instructors usually encourage team members to treat this stage as routine. Indeed, it is often difficult to tell the difference between a calibrated and an uncalibrated image visually. Emphasis on routine maintains codelike human tasks, as discussed above. For example, when I asked another PCC member how she thought the values on the graph were generated, she said, "I don't know why. . . . They're asking me to do it so I learn how to do it."[38] But this emphasis on executing the routine can also produce what members recognize as a "nominal anomaly."[39] For example, I once noticed some missing values in my graph and spent considerable extra time tracing which images had data loss and noting that loss in the log. A senior team member in the room discouraged me from doing what I thought was the responsible thing, insisting that it was not worth wasting my time on such a trivial error (or, more important, the time of the senior members of the team who would read my report and need to review the files). I wondered at the time why this might be. Was it just that this team member was tired after a long session of calibrating, late for class, or a poor calibrator? The last option seemed especially unlikely because of the person's extensive coding experience and long history with the team. I noted the problem in my log anyway but soon received a reply from the head of the PCC indicating that this was a known error and thus not really a problem after all. The next time I saw the same irregularity, I too could dismiss it as a nominal anomaly.

This incident demonstrates the acquisition of expertise as a calibrator. Adherence to routine does not necessarily indicate laziness or ignorance. Just like knowing what computers and humans are good at doing, knowing just what degree of precision in image maintenance is required, and how much is undue attention to unimportant details, is essential members' knowledge. It arises from practice with calibration materials and guided experience through group mentorship. Thus routineness itself becomes one of the virtues of calibration, since it demonstrates expertise with the system and its nuances.

Routineness also builds up the kind of expert fluency that permits the calibrator to recognize a problem—"strange stuff"—if it ever does arise. For example, a few weeks later I noticed another problem with my calibration procedure. The incident above prompted me to report this error as well rather than dismissing it with false confidence or for fear of alarming my superiors: I was thanked for noticing a "real" problem. Accumulating knowledge of "nominal anomalies" also establishes the boundaries of normality in calibration so that the routine can continue to be completed with regularity and team supervisors can focus on new errors instead of known problems.

Calibrators, then, aim toward an ideal behavior that is both tightly circumscribed and expert in terms of skilled members' talk and action. The circumscription of calibrators' activities is not a case of labor deskilling, as is frequently of interest in sociological literature on technology in the workplace; nor am I arguing that calibrators ought to know more about the code or be more involved in how it works. After all, the thought experiment suggested by the counterfactual—a hypothetical world in which undergraduates were allowed free rein over calibration procedures—would certainly undermine the perceived trustworthiness of the calibrated images. Rather, local judgments of efficacy and trustworthiness are predicated on precisely this careful balance of limitation and expertise. A trustworthy calibrator, like a trustworthy image, is a disciplined one.

Calibration is therefore another place where we might witness the crafting of research subjects at the same time as the crafting of research objects.[40] As they produce trustworthy images for the Rover team to use, the calibrators are disciplined too. The members' narrative is one of negotiation between humans and machines, of humans exerting judgment over computational scripts. But members' practices involve training human operators themselves to be as standardized and interchangeable as code or machines. Thus the calibrator's preserved naïveté is itself an important aspect of generating trustworthy instrumental reports. While digital images are constantly under manipulation, appeal to both the expert yet impassive human and the impassive code in the calibration process is central to generating and cementing trust in visual results.

Conclusion: Calibration and Digital Objectivity

Should all go well, as it regularly does, the calibrator has only to enter a few more small commands in order to upload the corrected images to JPL and e-mail a copy of the report text file to the Pancam team. Calibrators see "Done!" on their screens, at which they usually breathe a sigh of relief and gather their things to race to an impending class. The entire process, run side by side for both rovers, takes up to four hours. But mission scientists describe this time commitment as essential to providing them with trustworthy images of Mars. The camera's data results have been disciplined to approximate how the camera itself might have behaved had it been adjusted for local conditions. These local conditions are not the planetary geologist's concern: they are considered artifacts that must be removed from the data, or at least aspects of the planet that must be tamed or factored out in order to do science on Mars. Now that such distractions are drawn out of the image, scientific analysis can begin.[41]

In the face of the malleability of digital images, calibration is one way to ensure that the image is drawn as something credible, a trustworthy representation of the Martian surface. The image data is *drawn as* tamed and disciplined, stripped of unwanted characteristics while heightening valued features. At the same time, how the digital image is calibrated reveals members' fundamental assumptions about the proper role of observers, instruments, and computation in producing trustworthy observations. Focusing on the specific practices of digital calibration—of producing a trustworthy digital image for scientific analysis— reveals a prevalent local account of the moral and epistemic division of labor between humans and machines.

An instructive touchstone for analysis here is recent work by historians of science Lorraine Daston and Peter Galison, who have proposed that exactly how this labor is divided between humans and instruments points to a historically nuanced understanding of the notion of objectivity. For the nineteenth century, Daston and Galison describe what they call mechanical objectivity, in which credence for trustworthiness is accorded to graphing machines, such as the photograph or the phonograph. Observing scientists restrained themselves from intervening in the experiment and stepped out of the way of these impartial recording devices. The authors then argue that the twentieth century saw a rise of judgment in objective reports, in which the observer's experienced eye was required to interpret mechanical outputs, bringing out relevant features, or to explain how a singular case fit into a general scheme. In both cases the responsibility accorded to instruments and to observers in the process of knowledge production is clearly demarcated, and scientists at the time knew exactly how they must behave to guarantee trustworthy, objective results.[42] Which processes are at play in this twenty-first-century example, which enrolls both instruments and computation in scientific observation?

On the Rover mission, humans and software scripts work together in a hybrid, locally accountable way to draw the images that return from Mars as trustworthy visions of the Red Planet. The software scripts that transform the images one by one adjust each image automatically, so that they conform to the Rover scientists' ideals of what makes an image reliable, data-worthy, or otherwise scientific. That ideal is one in which the situated nature of the image must be effaced: identifying and removing local conditions makes images that are translatable across locations, times, viewers, and contexts on Mars. Further, the code digitally alters image data at the level of the pixel so that the resulting calibrated image possesses all the virtues of a trustworthy data point. The result is an image that is changed so one can almost believe the camera itself had been physically adjusted before it began to record images of phenomena. But the result is also an image that encodes in its very pixel composition what it means to be "calibrated." Thus individual images are drawn as trustworthy, standardized, comparable datasets, a view from nowhere produced by a computational modest witness.[43]

At the same time as this story relies on computational scripts to tame image data, it also integrates human judgment to tame the computer's too-perfect eye. Calibrators are called on to intervene and exercise their visual judgment at various stages of the camera calibration routine, even as computers are called on to reproduce routinized scripts. Inasmuch as this produces a view that Rover scientists consider trustworthy, it simultaneously produces a local understanding of what humans and machines are each "good at doing." So this story of digital objectivity is both hands-off and hands-on. It enrolls trained eyes, human error, and checks on impartial circuit arrays to produce trust in images, as well as software scripts that impress a preconceived notion of objective vision onto the image itself. And even as human judgment is exerted, mechanical qualities remain, whether in the black-box software, the carefully scripted actions of the human calibrator, or even the robotic body the cameras are mounted on. The mechanical, the computational, and the human cannot be so easily distinguished in practice, but at the same time, the unique properties and contributions of each of these actors are carefully demarcated in calibrators' accounts.

Despite the apparent impassivity of the camera's eye on Mars, then, the image's force as a trustworthy account cannot in fact be effected if "all sources of persuasion seem to have disappeared."[44] Because calibration is so essential to the trustworthiness of the data, the story of calibration must be told and re-told. After all, it is at this stage in image making—the calibration stage—that appeals are made to standardization, to cleanliness, to impassivity, and to the eradication of observer bias through visual technique. It is also in members' accounts of calibration that we see ideas and ideals of digital objectivity most clearly expressed: a particular configuration of humans and computers working in concert to produce a trusted view of the planet. The very idea of a seamless vision of the Martian terrain is not due to the camera's eye alone. Instead, the continuous taming of the observational field, effected by corrective software and human judgment, produces trust in instrumental reports and prepares images of Mars for scientific analysis.[45]

Image Processing
Drawing As and Its Consequences

The SOWG meeting has just ended, and the chatter of the Rover teleconference line is muted by the gentle sounds of classical music, piped into the office via satellite radio. I am sitting with Ben at his desk at a US Geological Survey branch office.[1] A Rover science team member, Ben has worked at the USGS for many years as a planetary scientist trained in geology. He is peering intently at a Pancam image of a rock at the edge of Victoria Crater that he recently requested that *Opportunity* photograph with the Pancams in thirteen filters (fig. 3.1). During the SOWG that planned the maneuver, the rock was given the target name Cercedilla. A black-and-white image of the rock is splayed across his dual screen display. To Ben, Cercedilla looks suspiciously like a piece of rock thrown outward from the deep innards of Victoria Crater during the impact that formed it: in geological terms, Cercedilla may be a piece of crater ejecta. If this is indeed the case, it would be useful for a geologist like Ben, who wants to know more about the deeper (and therefore older) layers of Mars that the crater's formation exposed.

Fortunately, *Opportunity* was commanded to take images of Cercedilla through each of its thirteen filters. Over the next two hours of

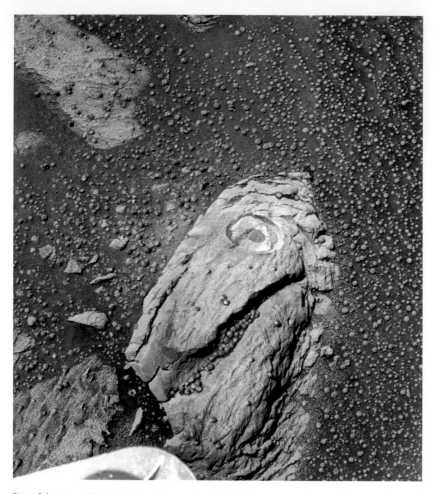

Figure 3.1. Cercedilla, single-filter Pancam view, filter L5. *Opportunity* sol 1184. Courtesy of NASA/ JPL/Cornell.

digital work, then, Ben will use these images to "characterize" Cercedilla: that is, to analyze its geological characteristics. To do so, he will compose and recompose the image of Cercedilla into various visual forms. Ben's work will involve software tools, screen work, gesture, and talk as ways of making sense of the digital image on his screen.[2] But all his work will also actively disambiguate, at each click, an otherwise ambiguous image. As an observer, then, Ben is not passive: he actively composes the image into something meaningful. The image that results records and embeds that legibility within its frame, so that the classifying, sorting out, and discriminating work of observation both arises from and is recorded in the work of digital image processing.

In this chapter I will describe the practical work of image processing—activities, forms of talk, interaction, imaging conventions, and instrumental techniques—that Rover scientists use to make sense of digital visual materials.[3] At Ben's desk, we will witness how interpretation, skilled vision, and the expert work of discriminating between kinds of objects are crafted into and through scientific images as they are processed.[4] This is *drawing as* in action. And it is through these practices of *drawing as* that other scientists will come to see the object of interest in just the same way. Turning from Ben's work with Cercedilla back to Susan's work with Tyrone, I will discuss how the work of visual composition presents implications for further observations, representations, and interactions among members of the Mars Rover team.

Image Work and the Dawn of Aspect

To understand how work with images can be considered scientific, it is helpful to review the way planetary scientists characterize their cameras and the images these instruments produce.[5] Central to this story of digital imaging in planetary science is the digital photographic plate: the CCD, or charge-coupled device. Instead of a light-sensitive plate that changes color with exposure, scientists describe electrical detectors that precisely count the number of photons that hit them. The standard explanatory analogy is the water bucket: in this account, detectors sit passively like buckets, counting the drops of water (photons) that fall into them.[6] As the "buckets" are tallied up, the resulting numerical value is expressed as a pixel value. This pixel data can be displayed either as a number or as a value of a shade of gray in a spectrum from black (zero photons) to white (many). As other analysts of digital imaging have described it, then, the digital image is both pictorial and numerical.[7]

According to Rover scientists, CCD-collected pixels represent both photon quantity and quality. When paired with optical filters, pixel values reveal information about an imaged object's ability to reflect light in a particular wavelength. This can be used as a diagnostic tool to identify mineralogical composition. As raw data, each individual image frame just looks like a black-and-white picture in which each pixel corresponds to the number of photons collected through the filter of choice. But combining these filtered images in an image processor through red, green, and blue data channels produces varying color images of the Martian landscape. Because the more extreme colors are produced by wider disparities

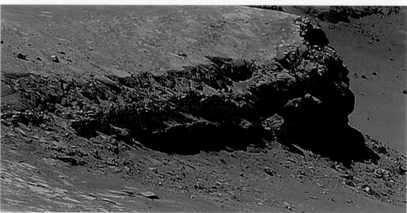

Figure 3.2A. Pancam observation of Cape Verde assembled in L257 false color. *Opportunity* sol 952. Courtesy of NASA/JPL/Cornell.

Figure 3.2B. Pancam observation of Cape Verde assembled in Approximate True Color with adjusted contrast. *Opportunity* sol 952. Courtesy of NASA/JPL/Cornell.

in pixel values between the filtered images across these channels, the resulting colors are taken to be clues to the object's chemical and mineralogical qualities.

On the Rover mission, the Pancam's thirteen carefully chosen color filters enable the team to take many filtered images of Mars from the same camera angle. These filter sets are considered particularly useful for seeing particular

kinds of features and are often combined and recombined in the course of mission operations depending on which features individual scientists most want to see. For example, a soil scientist interested in the composition of the terrain of Cape Verde, a promontory on Victoria Crater, assembled the left Pancam second, fifth, and seventh filters (abbreviated L257) in false color; this combination was judged helpful for revealing a wide range of textural and compositional differences (fig. 3.2A). The resulting picture was well received by soil scientists and doubled as a good image for planning a drive into the crater, since it highlighted different types of soil that might be hazardous or safe for rover wheels. But another geologist pointed to the same transformed image and said: "We think we're getting all this [great data], but look, what do we get [points to shadowed region]? Artifact soup." This scientist was most interested in characterizing the crater's stratigraphy: for him, "lighting and geometry" were more important than compositional difference, since they would allow him to measure the exact shapes, sizes, and depths of the crevices on the cliff face. He therefore combined the filtered frames that showed the least variation in pixel values and adjusted the lighting saturation to better reveal these distinctions (fig. 3.2B).

In these two renderings of the same image we see a switch between the artifact and the object of scientific analysis: composition and texture at the expense of lighting, or stratigraphy at the expense of composition. The pair of images also demonstrates how the selection and combination of raw images varies based on the image processor's intent: exactly what they want to see. But the flexibility to see it both ways is crucial to the science and operations of the mission. The geologist would not be satisfied with the soil scientist's picture, and a rover driver could not hope to identify slippery soil in the geologist's image. Both representations were derived from exactly the same dataset, the same set of pixels, but as a result of the choices of the image processor, a different set of features is revealed or subdued each time. The result of this plethora of possibilities is that one is often confronted with an image of an object on Mars repeated through different filters or processing algorithms. With so many possible viewings, it is clear that there is no one best way of picturing Mars. Rather, such images represent different ways of seeing and knowing the Martian surface.

In fact, the key to understanding rover images is that they are never singular views. Image processors combine multiple images over and over again to craft new visualizations of Mars. This is not a response to resource scarcity. Rather, the Rover scientists I studied explain that it is always necessary to see different things in the same image. For example, as discussed in chapter 2, when calibrating images that return from the panoramic cameras a human operator works in tandem with the computer to locate and eliminate light pollution, scattering, and dust across Pancam images. The resulting equation is applied across the board to an entire suite of images to systematically subtract a value from all pixels so that the images are corrected for dust and atmospheric conditions on any given day. But one person's artifact is another's data: many of the atmospheric scientists rely on these dust values to understand the atmosphere and Martian weather patterns, and soil scientists try to understand the optical quality of the dust itself. They therefore use the output from the calibration procedure to get the dust information and would rather see the dust than the image it obscures.[8]

The multiple views that result are therefore not an attempt to home in on a better representation of Mars in some absolute sense, or to produce incommensurable representations of the planet. Instead, digital image-processing techniques enable a switch between the artifact and the object of scientific analysis: composition and texture at the expense of lighting, or stratigraphy at the expense of composition. This ability to see the same visual data in different ways recalls the famous phrase *seeing as*, proposed by philosopher Ludwig Wittgenstein in the mid-twentieth century.

Wittgenstein illustrates *seeing as* with optical illusions involving ambiguous pictures, called gestalt figures in psychology, such as the duck/rabbit, the profiles/trophy, or the old woman/young woman pairs (fig. 3.3). He notes that people

Figure 3.3. The duck/rabbit. Jastrow, "Mind's Eye," 312.

do not usually say "I see it as" about their visual experiences—they just see. But the ability to say, "I see it as" arises in situations where there is some ambiguity about which features are salient: which elements form the background and which the foreground. This is the case with the gestalt figures. While the image does not change, in appreciating its same components in a different way you may suddenly experience a different observation, where the foreground and the background, or the artifact and the object, shift. This is when people stop saying, "I *see* a duck" and start to say, "I *see* it *as* a duck." Wittgenstein calls this moment "the dawning of aspect": a change in the organization of visual experience. Although the object does not change, this change of aspect produces a different observation, "quite as if the object had altered before my eyes."[9]

Like the duck/rabbit, we might *see* Cape Verde *as* a stratified cliff face or *see* it *as* composed of different soils. Unlike Wittgenstein's examples, however, these *seeing as* experiences are not "found" but crafted, the result of image processing. These actions and interactions compose the image into something meaningful, distinguishing foreground from background and object from artifact. Thus an interpretation or skilled vision is crafted into the image from the outset, so that the resulting picture incorporates elements of what it ought to be *seen as*. This is the work of *drawing as*.

Making It Pop Out

To witness *drawing as* in action, let's return to Ben's desk at the USGS, where he is squinting at the image of Cercedilla on his screen. With each transformation of the image, Ben attempts to disambiguate the visual experience of Cercedilla by isolating a single aspect of it at a time, blinding or curtailing alternative aspects. He purposefully includes particular features that he considers salient and simultaneously excludes or silences other features, relegating them to the background.

For example, one way to see Mars is by combining a set of filters through red, green, and blue channels in an image-processing program,[10] producing what the Rover team calls an Approximate True Color (ATC) image: "an estimate of the actual colors you would see if you were there on Mars."[11] This does not mean that true color images are any more "true" than other kinds of images. It is a technical term that refers to a particular combination of filters that approximates the range and type of light sensitivity exemplified by the human eye. The result is a Mars that looks reddish brown.

Ben, however, is not interested in what a human eye could see on Mars. Instead, he is interested in seeing which parts of the rock reflect light differently, since this could be a clue to mineralogical composition. He therefore asks the computer to show him aspects of Mars that the human eye cannot see but the rover's filtered cameras can: the near-infrared region spectrum of light. He loads the Pancam image-processing software and selects several filtered frames of Cercedilla pictures from among the Pancam thirteen-filter set that bear no relation to the human eye's sensitivity. As he combines these images through red, green, and blue channels in his image-processing software, the image of Mars on his screen brightens with bright yellow, blue, and purple (fig. 3.4): a false color image.

False color, to the Rover scientists, does not imply a false image; nor is the image artificially painted to produce spectacular views. Rather, the colors arise from a mathematical relation between pixels across the included image frames, enabling the viewer to see when objects in the scene reflect light in different wavelengths. Thus the distribution of colors in a false color image demarcates, highlights, or otherwise identifies invisible features of the imaged terrain. As one graduate student I interviewed explained, pointing at a false color image that presented Martian thermal data, "That is something you *cannot* see, so it looks like something you *can* see."[12]

As Ben describes it, putting an image into false color like this brings out new features that are otherwise invisible. In false color, "a lot of these . . . rocks suddenly pop out that weren't there before." This kind of talk is not unique to Ben but is echoed across the mission. Rover scientists frequently explained to

Figure 3.4. Cercedilla in a false color view, from a combination of filters. Author's photo.

me that the point of generating false color or stretched images was "to see new things," or to make a hidden feature "pop out." One scientist I interviewed who was looking for sulfate content on Mars explained, "If you get a particular [filter] combination the sulfates just jump out at you. It's like they turn green or blue or something." This change of view does not imply a change in the underlying dataset, only a change in orientation or aspect. As another scientist explained to me, "The data is the same, the difference is in what you see." A Rover Planner on the team echoed this statement: "The image never changes, but you can manipulate the image, and everyone sees something different."[13]

Certain filter combinations have become conventional on the mission, since they are considered particularly good ways of seeing locally relevant details. The most common combination is L257: the left Pancam's second, fifth, and seventh filters. This combination, as the SOWG Chair explained to Cynthia and Alexa in chapter 1, gives a broad enough range of coverage across the visible spectrum to highlight spectrally distinct objects in the terrain. Other combinations are more

Figure 3.5. Blueberries, L257 false color view. *Opportunity* sol 42. Courtesy of NASA/ JPL/Cornell.

specific. For example, when *Opportunity* landed on Meridiani Planum, the rover was surrounded by small round marbles of hematite that the team now calls blueberries. The mineral hematite is often formed in aqueous environments and appears to the human eye as dusty gray stone. But because it is slightly less red than most of Mars, combining the images produced by the fifth and seventh filters on the right-eye Pancam (abbreviated "R5-R7") which tend more toward the infrared, makes the hematite light up bright blue in the resulting combined picture. Because this particular visual construal makes the blueberries "pop out," the team calls this combination "the blueberry finder" (fig. 3.5).

Cercedilla also appears to be covered with and surrounded by blueberries, but it is unclear to Ben whether these blueberries are embedded in the rock or sitting on top of it, windblown from across the Meridiani plain. If the rock is crater ejecta, the two possibilities present different likely geological histories for Victoria's deep interior, one involving water, the other not. Ben therefore investigates the blueberries even further to better understand their distribution. Taking his false color image, he heightens the contrast between the different filters, creating a *decorrelation stretch*. "Stretching" here is a technical term that refers to increasing the contrast between pixels, roughly analogous to using the "contrast" tool on Photoshop. In a decorrelation stretch, the scientist increases the contrast in at least one of the combined filtered images by a certain factor but does not necessarily apply the same factor of stretch across the board to the other images in the combination. This changes the "correlation" between the pixels across the image frames. As Ben manipulates the sliders on his screen, Cercedilla brightens as if painted by pop artist Andy Warhol (fig. 3.6). He exclaims, "If you look at it like this [stretched], wow! That's really a different color. Suddenly there's differences in what I thought were really the same [thing]."

Having identified these differences, Ben moves from simply discriminating between colored materials to characterizing them in order to say something about their classification or origin. To do this, he draws on another common approach to image processing: producing a *cube* (sometimes spelled "qub"). Image processors talk about combining filtered image frames almost as if they are creating a stack of semitransparent photographs, layering one on top of the other in perfect alignment. Looking top-down at this pile, the combination of filtered frames produces a colored picture. But looking at the pile from the side, they see individual pixels perfectly aligned, each with a different value. They therefore speak of "slicing through the image cube," generating a graph of pixel values (i.e., how many photons an object collects at a single point) for each filter. These graphs are considered diagnostic for mineralogical composition. Because

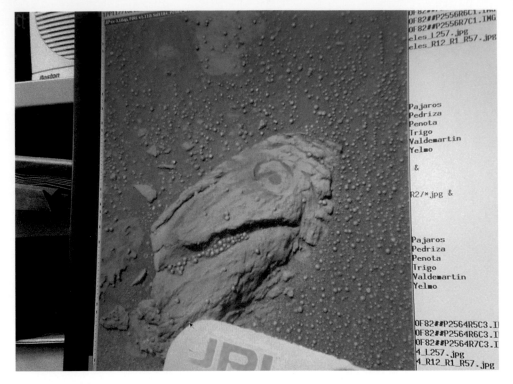

Figure 3.6. Cercedilla in a decorrelation stretch. Author's photo.

the object will absorb and reflect different quantities of light wavelengths depending on its particular mineralogical composition, scientists "read the spectral
signature"—the graph of pixel intensity at a single point collected through each
filter– to make claims about rock composition.

"Using the false colors as a guide," as he puts it, Ben starts by selecting an
area of the image of Cercedilla, toward the middle of the telltale circular stamped
depression left by the Rock Abrasion Tool as it ground into Cercedilla. The
software colors his selection red on the picture, and a graph pops up showing
thirteen red points connected by a red line (fig. 3.7). Ben peers at it. "Interesting," he says, and pauses. With his cursor he sweeps over the tail end of the graph.
"See the upturn? That's kind of blueberrylike. And it's from this center spot." He
moves his gaze and his cursor from the graph to the image, pointing to the swatch
of red. "So I'm gonna choose a different color and look at [he selects a region on
the edge of the RAT hole in green] that." Thirteen green points show up on the
graph alongside the red but do not follow the characteristic blueberry curve. He

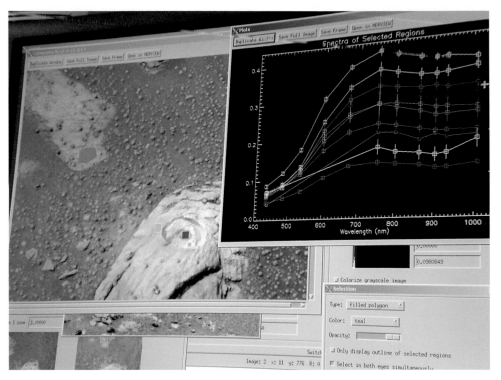

Figure 3.7. Using false colors as a guide, Ben selects areas on the image of Cercedilla to display a graph of pixel values across the thirteen Pancam filters. Author's photo.

gestures again with his mouse, sliding over first the green lines, then the red lines to point out the differences between them. "So there's the difference in spectra between the RAT hole [on Cercedilla] and a spot outside where the Mössbauer [instrument] got its data. And so, why are the spectra so different?"

Over the next hour, Ben transforms the same filtered set again and again. Each single aspect that the resulting image presents to view precludes other ways of seeing and knowing: the slope map image doesn't show him the band depth, the graph doesn't show him where the blueberries are located, and none of these images show him what Mars might look like to the human eye. As one of Ben's colleagues explained, "You have to throw out something in order to make it [the data] understandable." As each new aspect "pops out" I am reminded of Wittgenstein's description of the moment he came to see the duck in the duck/rabbit picture: when a change of aspect in how the elements of the visual field are organized produces a different observation, "quite as if the object had

altered before my eyes."[14] The ability to see the photometry or the blueberries is the product of skilled, disciplinary drawing practices that enforce an
aspect to organize visual experience and characterize the object in view. The
observer sees only the one aspect of the illustration along with the features that
the artist or scientist has determined are salient: what is drawn in, not what is
drawn out.

Analytically speaking, these are different *drawing as* practices, each producing
different possibilities for *seeing as*. It is as if the selection and composition of filters
takes the ambiguous duck/rabbit image and resolves it first into just the duck,
then into just the rabbit. However, the duck/rabbit example implies a particular
ambiguity in which there are only two possible ways of seeing the image. It is perhaps better in this case to consider examples that present many possibilities for
seeing as. Ludwig Wittgenstein uses the example of the "aspects" of a triangle in
this way:

This triangle can be seen as a triangular hole, as a solid, as a geometrical
drawing, as standing on its base, as hanging from its apex; as a mountain, as a
wedge, as an arrow or pointer, as an overturned object which is meant to stand
on the shorter side of the right angle, as a half parallelogram, and as various
other things. . . .

"You can think now of *this* now of *this* as you look at it, can regard it now
as *this* now as *this*, and then you will see it now *this* way, now *this*."[15]

Similarly, Rover team members describe images as concealing different kinds of information that talented image processors must work to reveal
by applying different visual conventions. As one explained it, "There's all
kinds of information in there [in the image]. These blueberries—it's not so
evident that they're made of such different material as the rocks they're sitting on. . . . Ross and Gwen [two mission scientists] really find some hidden
mineralogy."[16]

In this account, "information" is "in" the image. A cursory glance or even
a single-filter image is not enough to make distinctions in material composition
"evident." It is the skilled techniques of image processors like Ross and Gwen that
identify compositional distinctions by making them visible and observable. Click-

ing through decorrelation stretches on the Pancam image processor, this same team member described one image as "almost like seeing through the dust," while another "would not reveal . . . that we'd gotten into something different there." One decorrelation stretch was deemed more useful than the other because "[you shouldn't] waste your time on dust when what you wanted [to see] is the rock."[17] Which filtered images are combined and how they are displayed vary based on the image processors' intent: what they want to show.[18] Purposeful image construal, then, relies on mastering visual techniques that reveal certain aspects and conceal others.

Like lab work or fieldwork, it takes skilled membership to produce these observations. Ben analyzes Cercedilla using the masterful application of techniques that enable him to see the kinds of things geologists prefer to examine—mineralogical composition, texture, morphology, and so forth—and to display them in locally sensible ways. This kind of work is a primary component of scientific work on the Rover team. Most of the mission's scientists are trained geologists, geochemists, or atmospheric scientists, professions with a strong emphasis on fieldwork and lab work alongside computational analysis of datasets. But given their considerable distance from their field site, scientists frequently rely on imagery and image-processing software tools to produce knowledge about Mars. Digital image processing, to a large extent, constitutes the essence of "doing science" on another planet. Ben's colleague, Julie, concurred, telling me, "We [planetary scientists] have all become what they call 'pixel pushers' instead of field geologists."[19]

Such work has even colored, so to speak, how the scientists approach their fieldwork more generally. As Ben explained to me, gesturing to the bright hues of Cercedilla, "This is my fieldwork these days, and I sort of get used to the fact that this is the data you have to work with. I would almost feel frustrated being in the field and not having Pancam!"[20] Ben's eyes are fine-tuned instruments for fieldwork on Earth, thanks to his training in geology. Now he considers them deficient compared with rover vision. Other scientists across the mission frequently emphasized to me that the rover had the advantage of being able to see in different wavelengths than the human eye. Jude, a Pancam operator, described this as a feature of robotic space exploration more generally. As she put it, "We would not expect to see this [feature] without our instruments. That's one of the advantages robots have over humans."[21] But it is not only the robots or the instruments themselves that enable this kind of vision. A human with Pancam eyes would be limited to seeing through one filter at a time. Equipped with computational image-processing tools, the possibilities for seeing expand. It is not just the ability to "see in the infrared" that makes digital image work advantageous; it is the many visual combinations of a variety of filtered images, each presenting new aspects of Mars to human view.

Tyrone: Decorrelation Stretch

L5-L7-L2   **R2-R3-R7**

Figure 3.8. Tyrone, decorrelation stretch, from Susan's presentation. Used with permission.

From *Drawing As* to *Seeing As*: The Case of Tyrone

The true power of *drawing as* lies beyond the desktop: it is in interactions with other scientists that we witness an iterative relationship between these local representational practices and collective or shared seeing experiences. For an example, I return to this book's opening vignette: Susan and the case of Tyrone.[22]

A staff scientist at a midwestern research university associated with the Rover mission, Susan was a geophysicist by training when she joined the mission, but later she chose to complement her work on the rover's spectrometers with the Pancam's imaging capabilities. As she put it in our interview, "You shouldn't limit yourself to one [rover] instrument: it's the most foolish thing you can do!" During the Martian winter in which *Spirit* remained stationary, without enough solar power to drive, Susan traveled to a different university to spend time with the Pancam operators there, to train for the role of Pancam Downlink Lead, which requires reporting daily on the status of the remote instrument, and to learn to use the Pancam image-processing tools. While training, she practiced these techniques with recently acquired images, including the pictures of the patch of roughed-up soil at Tyrone (fig. I.2). As Susan recalls, it was while she was making false color composites that she first noticed that what looked like just a patch of white soil in a single-filter image produced different colors when composed into a decorrelation stretch (fig. 3.8).

Figure 3.9. A histogram of Tyrone pixel values (right), as seen on Susan's screen as part of her image-processing practices. Author's photo.

Intrigued by how something that looked like a single feature could perhaps be made of different types of material, Susan first turned to the numerical side of the image in order to characterize what she saw in the false color image. This would help her isolate the spectral properties of the two kinds of soils and possibly determine their composition. As she explained, "I'm not looking at a pretty image. I use [a] histogram . . . if my purpose [is] to see if [it is] two different types [of] material." Instead of asking the computer to generate a graph for a particular region of the image as Ben did, Susan asked the computer to display all the pixel values at once on a graph (fig. 3.9). That is, she *drew* Tyrone *as* a histogram: a graph in which individual pixel values are plotted together. Construed in this way, the image data showed two distinct clusters of pixel values. Susan interpreted these two branches of the histogram as different types of material, whose properties of light absorption were so different that they produced radically different pixel values in the image at hand.

Figure 3.10. Coloring in one branch of the histogram on the right screen in yellow and another in green reveals two distinct types of soil at different depths in the image on the left screen. Author's photo.

The Tyrone histogram showed that two kinds of material were present in the image data, but it did not show where that material was located or why it was changing. Susan therefore used another Pancam tool to "separate them [the two materials] spatially." When she colored in one branch of the histogram in green, all the pixels plotted on that branch lit up in green on the picture version of the same file. She could then see where that material was scattered. She proceeded to color the other branch of the histogram in yellow, lighting up a different patch of white soil (fig. 3.10). Thus two kinds of soil with different spectral characteristics were confirmed. And because of where those different patches of soil lit up in the image in green and yellow—what Susan called "spatial correlation"—she could tell that the yellow material was buried deeper in the wheel track than the green. Applying the same techniques to a series of images of Tyrone taken over several days, Susan noted that the histogram changed; that the yellow branch started to conflate with the green one (fig. 3.11). This suggested to her that the

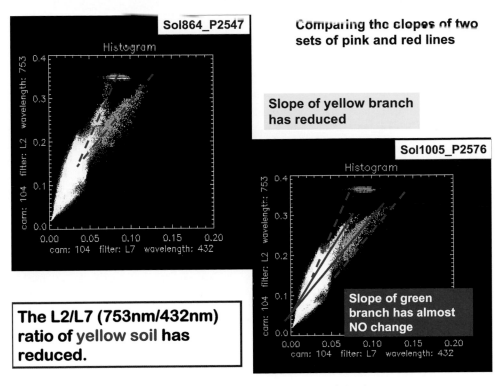

Figure 3.11. The slopes of the histogram change over time. From Susan's October presentation. Used with permission.

yellow material was changing in some way to become more like the green, perhaps owing to its recent and unexpected exposure to the Martian atmosphere.

So far this story is not unlike Ben's. As Susan *draws* Tyrone *as* a histogram, then *as* composed of two kinds of soil, her processing techniques reveal an aspect of organizing visual experience; and bringing several of these aspects together in concert, she makes a claim about a particular region of Mars. Each of these transformations also allows her to make an interpretative claim not just about evidence for two-toned material, but about its location and other characteristics. Where the story takes a novel turn, however, is when Susan left her screen to present this work to her fellow Mars Rover scientists.

Susan began by presenting these results to her colleagues at the End of Sol meeting in October 2006, the teleconferenced weekly meeting geared toward the presentation of ongoing, preliminary science results. She then requested further images of the Tyrone region over the Martian winter, while *Spirit* was stationary, and her fellow team members were suitably convinced from her presentation that they should include her "Christmas wishing list [sic]" of follow-up observations in Spirit's plan over the following week. A few months later she presented the results of these wished-for images at the face-to-face team meeting in February 2007, at Caltech in Pasadena, California. She began her presentation by showing the decorrelation stretch of Tyrone that she had first displayed a few months before. "You're all familiar with this beautiful Pancam image," she said. Then she applied the same stretch to eight pictures of rover tracks taken from across the region (fig. 3.12), narrating as follows:

> A similar situation happened in the Arad area, where we see the . . . color difference. This yellowish area shows this kind of spectra, and you have the slope at this kind of peak. . . . And when we do the decorrelation stretch we see the yellowish soil shows in the orangish in this area . . . also the purplish in the right eye is in the decorrelation stretch. . . . And at Paso Robles, we also see this area is the yellowish and the whitish [soil, in true color]. . . . At Wishing Well we also exposed some kind of lateral material. . . . we see there are also color differences.[23]

Applying the same visual convention from Tyrone to images taken across the region was a powerful representational technique. At this moment, the team came to see the two-toned soil, and see it everywhere.

But what could this observation mean? Susan next applied the same decorrelation stretch to Pancam images of Tyrone taken at different times in the mission. She showed that the histogram was changing slope, indicating a change in the properties of the white material. She cautioned, "We need to be sure this change is real, so I checked several factors." She next reviewed and dismissed the effects of a "diffuse sky" on how "the spectra behave," and any possible relation to optical effects of the camera using calibration data. Certain now of "the basic phenomenon of this observation," Susan suggested a change caused by atmospheric exposure and the subsequent dehydration of the salt properties of the soil. She corroborated this hypothesis with an experiment in her laboratory on Earth, showing that ferric sulfates decreased in acidity and could have affected the detected histogram slope.[24] Then Susan presented a topographical map of the area around Tyrone, highlighting the geographical locations of the observed light-toned soil. Considering that the soil was consistently visible in local lowlands, she put forward the potential hypothesis that this ferric sulfate deposit could have been distributed evenly throughout the region by something like flowing water.

WishingWell: Sol 351 - P2588-13F

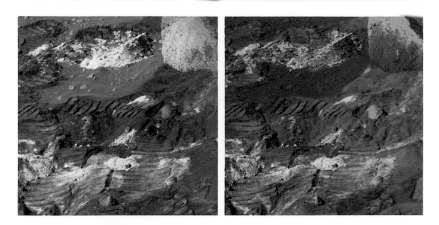

False color   Deco-stretch: L5-L7-L2

Arad: Decorrelation stretch

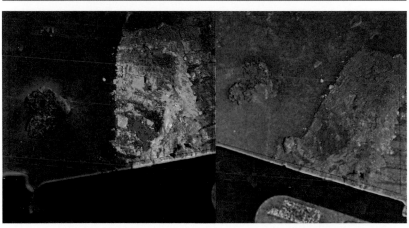

L5-L7-L2   R2-R3-R7

Figures 3.12A, B. Susan applies the same decorrelation stretch that she performed on Tyrone to other regions in her February presentation. Used with permission.

Susan's presentation was catalytic. The entire Rover science team began excitedly exchanging ideas and hypotheses about the light-toned soil. One scientist stated that "these observations make a compelling case" for some form of liquid water transport system in the deposit of the soils; another asked whether wind, instead of water, could have achieved the same distribution. Another scientist wondered whether volcanic processes could be responsible for laying down the salty deposits because of their high sulfur content, while another put up a slide showing an image of an environment in Iceland that she suggested "might be more consistent with what we're seeing" at *Spirit*'s site on Mars. Several other scientists took up the discussion of how old the salty deposits might be, with estimates ranging from millions to billions of years. The Principal Investigator extended the discussion past the projected end of the meeting to accommodate further conversation as scientists exchanged potential formation scenarios and raised challenges to each other's explanations.

All present treated the existence of the two-toned soil and its distribution as fact. The question up for discussion was not whether "the basic phenomenon of this observation" (as Susan put it) existed or how best to see it, but why it was there and how it got there. Discussion thus centered on different hypotheses about its origin and depositional mechanisms and generated proposed observations with the rover's suite of instruments to determine which of these hypotheses might be ruled out and which might be feasible or worth pursuing. When the discussion was summarized at a subsequent meeting, it was dubbed "the Light Soil Campaign" and encompassed a variety of observations aimed at better characterizing the two-toned soil at Tyrone and elsewhere. These observations formed the basis of rover operations for the following two weeks, and follow-up investigations on light-toned nodules that were also requested as part of the campaign formed the crux of *Spirit*'s investigations on the western edge of Home Plate.

Susan was adamant that the use of yellow and green colors revealed a distinction in the soil instead of adding or coloring in an interpretation. "The change was real," she said. But her use of color was important for "showing" this distinction both to herself and to others. Her initial interest in the light and dark rocks on the mission made clear the importance of conscripting other scientists to her point of view. With respect to Tyrone, as she put it to me when I visited her laboratory: "You decide the color you want to show, the color you want to use, but the data is there, it's not the color. . . . Because the existing data [images] contain this kind of information, you decide how you want to show [the data]."

Green and yellow thus became convenient ways of reconfiguring the pictorial representation of the image so that this feature of the soil "lit up" (or "popped out"). The colors also depict "information" that is "contain[ed]" *in* the image, not glossed onto it in interpretative annotations. This is important to team members, who distinguish between annotations as interpretations (discussed in the next two chapters) versus image-processing work that presents existing distinctions in the data. But while the image "contains this kind of information" (the spectral properties of the soil), it is at Susan's discretion to "decide how to show" the data. That is, *drawing as* practices allowed her both to see a distinction in the soil and to show her colleagues what to see in the soil too. Reconfiguring the soil in this way means that every time scientists look at the image of Tyrone, they see the two-toned white soil. Once the distinction has been made in one aspect, it cannot be unseen.

This is not limited to Susan's transformations of Tyrone, or to Pancam imagery alone. Across the mission, team members articulate the Wittgensteinian dawning of aspect when presented with a digital image that has been drawn so as to present particular properties. Expressions such as "now I see!" can be heard in SOWG, End of Sol, and science team meetings as well as at scientists' desks as they go through different image-processing routines or present these interpreted image products to their colleagues. As one scientist examined an image produced in his lab, he noted, "It's efficient to have something like that [image] to communicate what you're showing, what your interpretation [is]." Even in operating the MiniTES thermal spectrometer, a team member explained that he had to "show other spectra to teach [the team] what to see," or that he took the approach of "I'm only gonna show you the part I want you to pay attention to." This is not hiding data that might be essential to interpretation, but rather limiting data to the relevant part: an attempt to *draw as*, to delimit aspect in order to produce and reproduce a *seeing as* experience across the team. As I will show in the following chapters, this use of purposeful image construal to direct a viewer's attention in turn presents implications for the kinds of science and operations that are eventually planned as a result of collective visual interpretation. As a MiniTES operator explained to me, "the science questions come out of the imagery."

Drawing As, Seeing As, and Social Formation

Drawing as, then, is not only a question of making epistemic distinctions and visualizing an object, it is also a question of drawing distinctions and unifications among subjects, of drawing actors together into different social configurations. Even while *seeing as* experiences are produced by *drawing as* practices at Susan's

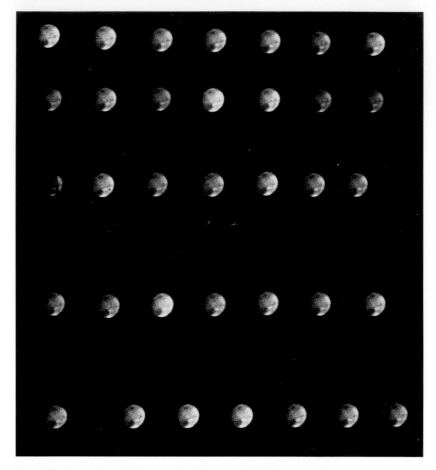

Figure 3.13.   Percival Lowell's photographs of Mars. Lowell Observatory Archives.

or Ben's desk, this seeing is social, intertwining both visual practices and social commitment.

Planetary images have long been complicit in this tight combination of *drawing as, seeing as,* and community formation. The astronomer Percival Lowell, well known for his insistence on Mars's canal network, battled the same issues of visual salience, expertise, and communication of categories. When in 1909 Lowell was invited to submit his photographs of the planet, taken through his famous telescope at his observatory in Flagstaff, Arizona (fig. 3.13), to the Dresden Photographic Exhibition in Germany, he initiated a long exchange with his colleagues Vesto Slipher and Carl Lampland about how to visually communicate what they could see.[25] Newly introduced to astronomy, photographs offered an unparalleled appeal to the public to see the Martian canals for themselves. But the scientists were aware that the photograph was itself ambiguous. Shades of light and dark played over the planet's surface, mechanically and passively inscribed, perhaps, but demonstrating precious little. Just presenting row on row of tiny photographs was not enough; the public had to be taught how to see them. Slipher therefore wrote to Lowell and Lampland: "What do you think should be placed along with the Mars Photographs in the way of drawings? To those who are not familiar with the difficulties in the way of success in such work (and they are 99.99%) the photographs might not come up to expectation if shown along-side drawings. . . . Now on the other hand, there must be something with the photographs to point out what to expect and look for in the photographs."[26]

To disambiguate the photograph and train the viewer in what to see, Slipher needed to *draw* Mars *as* a canal-crossed planet. One possible solution was to annotate the images by placing drawings next to the photographs, directing observers' attention to relevant features, parsing the photograph so that others could see.[27]

Nor is this phenomenon limited to photography. In 1609, Galileo Galilei famously turned his telescope toward the moon and produced one of the most famous images in the history of astronomy: a cratered, pockmarked moon (fig. 3.14). Historians of science hesitate to say that this drawing represents exactly what Galileo saw: we cannot know what image actually hit his retinal wall. But his drawing presents no ambiguity about what he presumes the dark patches on the moon to be. Using the then-novel technique of chiaroscuro (shape from shading), Galileo *drew* the moon *as* a topographical body, with craters and pockmarks.[28]

Note first of all that *drawing* the moon *as* a topographical body reveals where Galileo's theoretical commitments lie.[29] *Drawing* the moon *as* a topographical body makes a Copernican statement about what the moon is and how we should best understand it. The drawing need not be a perfect record of what Galileo saw, but the drawing is where the discovery emerges. The images in *Siderius Nuncius* present an excellent comparative example of how visual and theoretical insight is produced in and through the purposeful use of representational techniques and selectivity.[30]

But the case of Galileo is also important because it demonstrates the reciprocal relation between drawing and seeing. After a tour to the New World, where he had mapped the territory of Virginia, Queen Elizabeth I's geometer Thomas Harriot also turned his telescope toward the moon in 1609 and, presumably, drew what he saw: a crescent, some shading, and a dark patch near the center

Figure 3.14. Galileo's image of the moon, *Siderius Nuncius*, 10 C2R. By kind permission of the Institute for Advanced Study, Princeton, NJ.

(fig. 3.15). The image betrays little sense of what the moon is or how to organize this visual experience. Only a year later, in 1610, Harriot produced a radically different set of drawings of the moon, clearly emulating the recently published Galilean view: a pockmarked moon, divided perpendicularly into light and shade, with a giant crater in the center (fig. 3.16). Galileo's way of drawing the moon was a powerful way to communicate and reproduce his particular way of seeing—his skilled vision, his discrimination of categories, and his theoretical commitments too—even at a great distance.[31]

These historical examples make it clear that *drawing as* practices do not construct (only) the world on Mars or on the moon. They also construct communities on Earth. The case study of Susan and Tyrone is especially illuminating for how her *drawing as* practices translated into a *seeing as* experience that was taken up across the mission and that directed further rover operations. Similarly, the discussion of Galileo and Harriot shows how a depiction of the moon as a sublunary object both required and strengthened a community of astronomers who

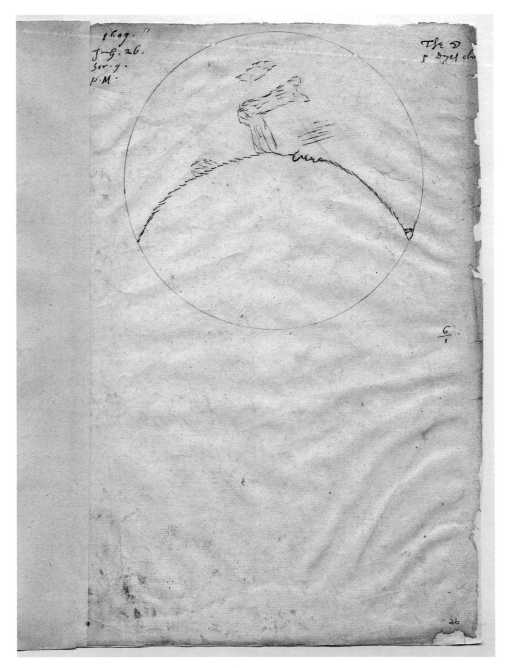

Figure 3.15. Harriot's image of the moon, July 26, 1609. Copyright Lord Egremont. Used with permission.

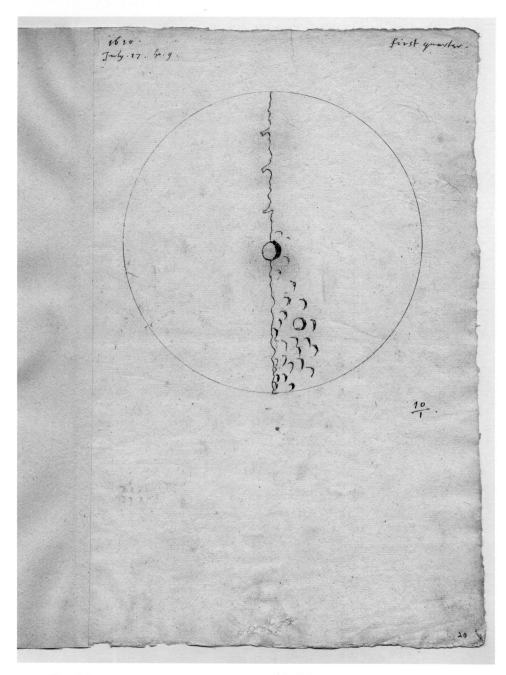

Figure 3.16. Harriot's image of the moon, July 17, 1610. Copyright Lord Egremont. Used with permission.

believed in the Copernican worldview and its practices, who took up Galileo's way of seeing and representing as their own. If seeing is social and *drawing as* practices produce and reproduce these modes of seeing, then how we represent Mars is not just a question of what *Mars* is like, or even of what we *think* Mars is like—it is about what *we* think Mars is like.

This is especially visible in the disciplinary heritage of *drawing as* practices. Scientists like Ben or Susan deploy representational practices shaped by their disciplinary training, but they may also appeal to those same disciplinary divides to ground their visual transformations or support their requests for particular images. Sam frequently explains his predilection for maintaining high resolution despite observation trimming as "I'm a geomorphologist, so I'll always take the higher-resolution image."[32] His colleague frequently prefaces his own graphs with "I'm gonna do a series of element to element diagrams—no surprise there, in that I'm a geochemist and all."[33] Even Susan, presenting her Pancam image results, joked to her peers, "I'm going to show a beautiful Pancam picture and pretend I'm a geologist."[34] These remarks establish a close relationship between disciplinary modes of inquiry and preferred visual forms.

Software packages, too, play a role in reifying these categories through practice, since they come preloaded with specific techniques for *drawing as*, presenting a ready-made *seeing as* experience to viewers consistent with disciplinary interest. Scientists use particular software packages specific to their disciplinary and institutional heritages: geographers prefer ARC-GIS or ENVI, the USGS developed ISIS for planetary studies, and astronomers use IDL. These different software packages implement differences in ways of seeing that can produce different aspects even when producing the same kind of visualization. For example, when Ben's colleague Ross produced a decorrelation stretch of Cercedilla using different software, the two images looked quite different. One could also see details in Ross's images that one could not see in Ben's, and vice versa.

As another example, different types of maps betray different disciplinary approaches and software tools. Tom's Geographic Information Systems (GIS) laboratory at a large state university creates Rover transit maps, while Peter produces Rover transit maps at NASA's Jet Propulsion Laboratory's image-processing center and Joseph uses orbital images to produce geological maps. The three use very different techniques. Locating the rover using orbital GIS data, as Tom does, versus using the robot's odometry, as Peter does, presents unique advantages and disadvantages depending on the slip of rover wheels or the availability of orbital coordinates. Tom described these different maps as a question of different disciplinary perspectives and expertise, produced through software suites and visual transformations: "Joseph looks at the images and interprets the rocks very well; he is a geologist. I am not, I'm an engineer. He's good at the tactical, we should go here, we should go there. That [my software team] can't do. He doesn't have the tools we have, the software. Peter, he's a geologist . . . he doesn't have the math models and software we have."[35]

Tom explained that what he described as his "software engineering" perspective on Mars had the advantage of mathematical modeling but the disadvantage of little geological interpretation, unlike Joseph's and Peter's. Each perspective is encoded in and produced through the different images. Software suites and visual conventions make some possibilities available, but they limit others by leaving them out of the picture. Drawing attention to different scientists' disciplinary heritages, such as geography or geology, chemistry or geomorphology, can demonstrate why their visual production presents so many different aspects of the same images to view.[36]

This emphasis on multiplicity, however, demonstrates only one aspect of the relationship between *drawing as* and social formation. Different communities may present their own unique practices, but on the Rover mission those visualizations are treated as commensurate—resulting in different but reconcilable visions of Mars. They are brought into coordination with each other through methods consistent with the team's consensus-based organizational structure.[37] The Rover team accounts for this practice with its native philosophy of science; as one team member put it, "When you see it in all these different ways, then you get to know it." In the following chapters I will describe these practices in more detail, showing how *drawing as* techniques that present disciplined ways of seeing are coordinated to produce singular views of Mars. In this way, image processing and the practices of *drawing as* produce not only scientific sight and insight, but scientific community as well.

Conclusion

Spirit's activities at the western edge of Home Plate cannot be understood without careful attention to Susan's representational work. First *drawing* Tyrone *as* composed of two distinct kinds of salty soils distributed at different vertical layers, and then *drawing* Arad, Paso Robles, and Wishing Well *as* Tyrone, encouraged the rest of the team to see Tyrone as composed of those materials as Susan suggested, and then to see other examples as cases of the same phenomenon. Following this work of sorting out distinctions through drawing and seeing, a suite of rover operations enacted the light-toned soil and brought it to light in each encountered location. Soon thereafter, published papers bearing Susan's name along with those of her Athena Science Team colleagues began to appear in the planetary science section of the *Journal of Geophysical Research* and in the prestigious *Science* magazine.[38]

Such activity arose as a result of specific practices of image processing that purposefully composed images of the soil so that the team could see what Susan saw. The interpretation is drawn into, inscribed in, and produced through the very images that present the phenomenon, such that the phenomenon can be seen. And as this visual convention was applied across Gusev Crater, the scientists no longer *saw* the white soil *as* two-toned: they simply saw the two-toned soil, and saw it everywhere. These are the activities that constitute scientific work with digital images. Practical work with images disambiguates visual material, shuts down ways of seeing in order to focus on one aspect, one set of salient relationships. These techniques of *drawing as* reveal and present different aspects with every click of the button, enabling different *seeing as* practices at the point of the observer, as in Ben's situation with Cercedilla. But they also powerfully transmit a *seeing as* experience to subsequent viewers.

Visualization in science, then, is not a question of creating an ever truer or more singular image of an object. Rather, it is a practical activity of *drawing* a natural object *as* an analytical object, inscribing a value into the very composition of what that object is and what makes it interesting, so that subsequent viewers and image makers will see, draw, and interact with that same object in the same way. Team members' mastery of image-processing software provides them with one of their most important strategies for materially realizing objects in the visual field. These are the techniques and processes of *drawing as*: the practical activities by means of which a *seeing as* experience is produced. And if *drawing as* can transform the subsequent *seeing as* experience into just seeing, then we arrive at the special power of the scientific image: that the drawn features of an object are *seen as* phenomenal or even ontological properties by the actors in question. That is, *drawing as* makes epistemology look like ontology. It conflates our interpretative work in the world with the objects we encounter there and draws them accordingly.[39] Tracing the practical actions of scientists engaged in purposeful visual construal, then, presents an opportunity to literally trace actors' commitments at play through an examination of both practical activities (of drawing) and practical effects (further representations and interactions). The scientific image itself does not so much document the object out there as document the work of different communities of knowing subjects that enable, produce, and constrain knowledge of the world.

"These Images Are Our Maps"
Drawing, Seeing, and Interacting

4

On the other side of the planet from Tyrone, *Opportunity* has spent almost a whole Earth year exploring Victoria Crater. Proceeding clockwise around the rim, the rover drives up to each ledge in turn and snaps high-resolution Pancam panoramas. In image after image, the gaping vista of the crater with its rippled dunes at the center opens beyond towering promontories and rock cobbles at the rim looking like dragon scales. "Who would ever have thought we'd ever take a picture of Mars that looks like that?" the PI gasps when yet another image of a cliff face comes down from the rover. He points at a photograph of dusty boulders imaged by the Viking lander, mounted on the wall in his lab. "Up until now, *that* was the most exciting view of Mars anyone had ever taken."[1]

The high-resolution Pancam images of the crater's capes are indeed a far cry from the dusty, rock-strewn vistas witnessed by previous missions like Viking or Pathfinder. But the team did not take these images for NASA's Image of the Week. They will use them to inform their decision about how the rover should circumnavigate Victoria Crater and where it should try to enter. Groups of scientists and their students will analyze the images for topographical information and try to get a geological sense of the regional environment. They will also import the photographs directly into the rover's command software to point the instruments and to plan new drives. The practices of *drawing as* therefore have implications not only for how objects are seen, but for how those objects are interacted with. As the PI once explained to me, "These images are our maps."[2]

Far from being accurate representations of a landscape, scholars of critical geography have argued, maps are the product of particular regimes of seeing and knowing.[3] Mapmakers' local distinctions, institutions, and commitments are projected onto the imaged terrain, naturalizing those distinctions and effacing their work of production. The mapmaker must delineate categories and kinds and negotiate both which elements to leave in and which to leave out. As such, mapmaking presents a particular form of *drawing as*.

But the power of maps is not merely in their making: it is also in their deployment. As Denis Wood puts it, maps work.[4] They work not only by naturalizing and operationalizing human knowledge and social structures, but also by presenting possibilities for our interactions with the mapped territory and suggesting potential for wayfinding within a space. This is especially true for the type of maps I discuss here. These are not official surveys by any means, nor are they released to the public. Rather, Rover maps are annotated rover images: marked up, circulated among team members, and used daily to structure conversations about how, where, and why to next interact with Mars. At most these maps may take a regional view by annotating an image from the MOC or HiRISE camera in orbit with inferences gained from interactions on the ground, again to guide driving decisions. Such maps not only are representations of the local terrain, they are propositions about how to intervene.[5]

In his discussion of the relation between representing and intervening, philosopher of science Ian Hacking argues that all representations are predicated on object interventions.[6] It is true that the rover must interact with Mars in some way to capture images of the planet's surface. But this chapter will reveal a more iterative relation between representing and intervening. I will describe the processes and practices for *drawing* images of Mars *as* maps that impress an interpretation onto the planet's landscape and direct rover interaction accordingly. This involves producing homegrown annotations, symbols, and names that identify, demarcate, and transmit ways of seeing to other members of the team. Occasionally it involves processing or compiling chains of these images into an aggregate view, achieved through either experiential or computational analysis, to visually present possibilities for rover activity. The representational question at stake here is not how well or even whether the visualization depicts objects on Mars, but how such visualizations construct Mars for robotic interaction.[7] Representational choices structure interventional choices too.

Holding our focus steady on the local practices and interactions with and around rover images, then, this chapter explains how and why scientists and engineers *draw* rover images *as* maps, and what kinds of knowledge and possibilities for interaction those maps represent. To do so, we must follow them into their strategic planning sessions: the weekly End of Sol meetings that host scientific presentations and long-term planning discussions. My first example is from *Spirit*, picking up the rover's story after the discovery of Tyrone, where I will show how images are enrolled in the production of scientific knowledge about the region by producing geological maps and by directing particular kinds of observations throughout the region called Home Plate. I then turn to *Opportunity*, where I will show first how scientists deploy annotations to preserve ongoing hypotheses about the region and then how scientists worked with three-dimensional imagery to inform the decision of driving to and around Victoria Crater. These stories took place over several months or even years of Rover work, and their telling here is brief, but I will return to them throughout the rest of the book. In describing how scientists use maps to put rover observations into context, I hope to provide some context for these observational exchanges as well. For the time being, I focus on the iterative relationship between *drawing as*, collective *seeing as*, and subsequent interactions. I refer readers to the team-produced Rover traverse maps in appendix B to ground these stories.

Mapping for Science: "Everything Is Colored according to Your Hypothesis"

First glimpsed at a distance from the top of Husband Hill, which *Spirit* climbed to escape the volcanic plains of its landing site, the Home Plate region earned its name because from orbit its shape recalled home plate on a baseball field. Home Plate's geological history was mysterious, but with the discovery of the two-toned soil at Tyrone, it became clear that the region had once been covered in water that could have distributed the white materials around the area before they were covered again with red Martian dust. One hypothesis following this discovery was that Home Plate could have been the site of a former hot spring, which could have produced minerals and deposited them over such a wide area. This would be a momentous discovery for the team, since geothermal hot spots on Earth often teem with microorganisms.

Hot springs feature not only hot water, but also soluble minerals in the water, like the salts at Tyrone. They therefore tend to leave behind thin, crusty layers of

these minerals deposited over time, as well as soft rocks built up as these minerals coalesce in the warm environment. So to establish whether Home Plate was
once a hot spring, the science team turned to characterizing the chemistry of the
soil and of the crumbly rocks scattered around the region and to describing the
extent and complexity of any visible layering. Over two years, the entire suite
of Athena science instruments was regularly deployed to build up evidence for
these claims: as the PI described it, "hitting Mars with everything we've got."
Spirit took hundreds of Pancam images of the sides of Home Plate and low
ridges around it, while a scientist and his graduate student at Caltech analyzed
those same images to determine everything they could about the layered deposits, how they were laid down, and what geological processes were responsible
for their layering. Every cobble in the area was subject to a Microscopic Imager
picture, MiniTES spectra, and, time permitting, Mössbauer and APXS measurements to fully characterize its chemical composition and textural features. *Spirit*
was even directed to drive over one such cobble to crush it under its wheel, and
then to take Pancam, MiniTES, MI, and spectral readings of its interior. Every
time *Spirit* drove, Pancam thirteen-filter images were requested of its wheel track
to detect silica in the area, and when *Spirit* stayed stationary for several weeks,
the wheel tracks were imaged again to see if any changes could be detected. Following these many months of observations in the region to the east of Home
Plate (which the team nicknamed Silica Valley), the team commanded *Spirit* to
drive on top of Home Plate to conduct more observations from there.

Each of these observations required considerable work to coordinate planning, calibrate images, and produce or display scientific results, as described in
the previous chapters. But these observations would remain discrete if not for an
additional type of visual work that drew the observations together into a coherent whole and grounded the region's scientific exploration.[8] Recall how a strong
point of Susan's presentation was her demonstrating that other regions showed
evidence of the same phenomenon she observed at Tyrone and then mapping
the location of those regions to show that they all occurred in relative lowland
areas. Coordinating *Spirit*'s many observations required a particular visualization, an evolving document that changed or expanded with the inclusion of new
local observations: a geological map.

Joseph, one of the Rover scientists and Long Term Planning group members, regularly produces geological maps for the mission. At a Team Meeting in
July 2007, he showed an iteration of this map of Home Plate (fig. 4.1), which
appeared as an orbital image of the entire region taken by the HiRISE camera
on the Mars Reconnaissance Orbiter.

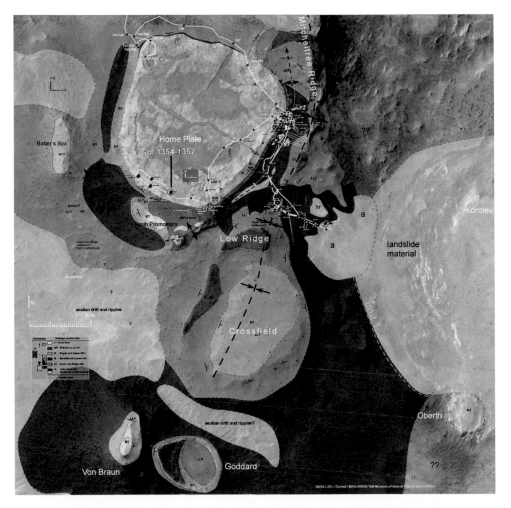

Figure 4.1. Joseph's "regional geosketch" map of the Home Plate region. Different colors represent units that are geochemically or stratigraphically similar. *Spirit*'s path is marked with a yellow line. This iteration is taken from an End of Sol meeting, October 30, 2007. Base image credit: NASA/JPL/University of Arizona. Used with permission.

The map I'm referring to is this simple geological sketch map of this area of Home Plate. . . . I'm gonna talk a little bit about what all of these colors are, discuss some of the ideas that went into building it, how I did it, and discuss some of the working hypotheses for the correlations and a little about some of the structural relationships that one might learn about by looking at how the geologic characteristics interact with the local [area].[9]

Joseph had used a digital paint program to color in this image with large patches of purple, pink, and yellow to indicate which areas at Home Plate he believes are related and which are distinct. Throughout his presentation, he went through several steps of visual parsing, calling the group's attention to features through his annotations. He referred to false color images of Tyrone and other features, as well as some of the chemical observations of local rocks, to indicate what discoveries had occurred where and how they correlated with other observations in the region. Unlike Susan's or Ben's image-processing techniques, the bright colors in Joseph's geological map did not arise from combinations of filters: they were his annotations on an orbital image as a base. His map ensured not only that his colleagues could identify the same features in the region, but also that they too would see them as salient and grasp the story he was trying to tell about the region.

This kind of drawing and related image work has been central to geology since its inception in the late eighteenth and early nineteenth centuries, as Martin Rudwick has documented.[10] Geological maps are one of the visual techniques exported from Earth to other planets under the interdisciplinary field of planetary science. Using an orbital or aerial photograph, the geologist synthesizes individual observations acquired on the ground into an interpretation of the site's general characteristics, then colors the image in various shades to represent different geological units, thereby imprinting his or her interpretations onto the landscape.[11] Such maps might identify categories like "regolith" or "contacts" between different geological units, for example. Maps of this type may also be subject to peer review and published by the US Geological Survey. However, beyond simply identifying areas, many Rover scientists impressed on me that a geological map must "tell a cohesive story" about that region. One cannot identify volcanic rocks right next to rocks formed with substantial wind erosion, for example, without some kind of geological narrative about why those two elements should be found so close together without influencing each other. Taken together, the annotations of different units or rocks should enable a geologist to identify the processes or periods of deposition—to build up a geologically robust narrative about the history of the area under scrutiny. Joseph therefore describes his geological maps as "a sort of X-ray vision version of the landscape in which everything is colored according to your hypothesis"[12]—a revealing statement for a discussion of *drawing as*.

"Coloring according to your hypothesis" not only records knowledge about a region, but also serves an ongoing purpose as grounding for active fieldwork, enabling scientists to keep track of hypotheses as they go along. Joseph, like other geologists on the mission, was classically trained to keep notebooks of evolving hypotheses about the region during his fieldwork. He is therefore careful to call his maps "sketch maps." Presenting one of these at a Team Meeting, he was clear about the iterative relationship between drawing and ongoing fieldwork:

> I used this word sketch map, and why didn't I simply just say geologic map? The idea is that the sketch map traditionally is something you sketch out in your field notes as you're working along, and it's basically the idea that you're doing the mapping and you're modifying it as you're actually doing it and you update your hypotheses. . . . It's a process of continually presenting to yourself hypotheses about what you think you're seeing and then testing immediately. It's a continuous process. So a sketch map is just one of those steps in that process where you map out what you think you're seeing, and then you move to test various contacts. . . . [It's] a field-based best estimate of geologic units at any given time . . . putting [your interpretations] into the base map and then building your geology around you.[13]

The geological sketch map presents and preserves evolving hypotheses about a region, placing discrete observations like piecing together a puzzle, to generate an overall picture of the region's geological history.[14]

On Earth, a geologist would construct a sketch map by walking around an area, frequently circling and returning to locations to establish their contents and relation to each other. As Sam, another Rover mission geologist, described it to me, "you have to walk the contacts" of a region; find the areas where there is some kind of distinction between one set of features and another.[15] On Mars, however, constructing a geological sketch map and testing its hypotheses requires using the rover's instruments to characterize different regions and reveal distinctions, then aggregating these discrete observations over time. The hypothesis that ties these observations together is one that indicates not only which areas are which type of geological unit, but also why: what specific geological history would have produced such a landscape with these particular observable features. The map also helps to identify where the rover needs to go next and what it must do there to fill in the blanks or further support the hypothesis. Joseph frequently places question marks on his map to show locations that require further analysis, and he describes the process as "testing" "what you think you're seeing." The iterative relationship between *drawing as* and *seeing as* endures in the practices of mapping as well.

Annotation, Attention, and Coordination

Geological maps are an effective way of synthesizing information about a region to be presented at a single sweep. But the process they represent is far from linear. In order to *draw* certain aspects of the region under study *as* similar or distinct, those aspects have to be seen as salient in the first place. In *Opportunity*'s investigations of Victoria Crater, we witness another form of image annotation used to direct the team's visual attention toward a feature that one scientist believed was salient.

At Victoria Crater, the Rover team was interested in a different scale than at *Spirit*'s site, in terms of both the size of the area under investigation and the extensive age of the region. Stratigraphy and geomorphology played a primary role. Geologists are especially interested in impact craters because they carve out deep holes in what is otherwise bedrock—the base layer of rock that is placed or altered over millions of years. Impact craters like Victoria expose these ancient layers to view. *Opportunity* had already visited a few smaller craters in the area before moving to Victoria, but it had not yet been to such a large, deep crater. The goal at Victoria, then, was to "drive around the crater and then go into it and measure the sections, . . . looking at the chemistry as a function of the stratigraphic level."[16] The team hoped that the lines and striations in the cliff faces of Victoria, as well as targeted spectral readings of each one, would give clues to how the layers of bedrock were built up in the first place, and therefore point to what kind of place Meridiani Planum was hundreds of millions of years ago.

Once *Opportunity* reached the crater rim, its cameras took several images of the promontories to the north, including one called Cape Verde, where the striations in the rock face were visible. Later in the circumnavigation of the crater, the Long Term Planning Lead outlined the tactical plan for a high-resolution Pancam imaging campaign of the rock face to try to identify those very striations, which would give an indication of the crater's possible formation mechanism. The LTP Lead began:

> [This slide] shows the sol 1002 Navcam, a pair of Navcams spliced together, and I've noted in a dashed or dotted yellow line there approximately the location where at least my eyes think I see this thin laminating unit which is one of the targets. But it's really this entire face of Cape Verde that we're trying to image. So with that in mind let's look at the [next] slide, and again this is just a reminder of the Cape Verde stratigraphy, looking back from the other side, and this is the thick bedded facies that we're really trying to image.[17]

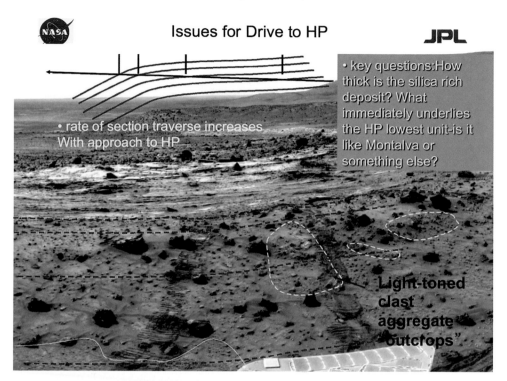

Figure 4.2. Using dashed lines to draw on a rover image to indicate distinctions between layers. This particular iteration was shown at an End of Sol meeting, May 21, 2007. Credit: NASA/JPL/Caltech. Used with permission.

The LTP Lead here uses dashed lines to indicate where "at least my eyes think I see" a stratigraphic layer of interest at the edge of Victoria Crater, thereby *drawing* the structure in front of the rover *as* a thin laminating unit (fig. 4.2). This does the work of capturing the discernment so that the vision can be shared, enabling the rest of the team to see the area as a laminating unit as well.[18] But it also does the work of suggesting what the structure is and how it came about. Thin laminating units do not appear out of nowhere; they require a particular geological history to develop into what is visible today. Following the Navcam image display, the LTP Lead included a "reminder" of the stratigraphic context the team was trying to elaborate at this location and the "thick bedded facies that we're really trying to image." Simply directing a colleague's attention to an element in the scene is not enough: one must also offer an analytical reason this object in the scene ought to be noticed or considered salient in the first place. Using the technique on another image of "some rather interesting near field surface textures" on the far wall of Victoria Crater, another scientist described this annotation style as "sort of suggestions that are encoded into this image."[19]

Drawing on the image, as sociologists of science Karin Knorr-Cetina and Klaus Amman have shown, is an important way of discerning salient features for display to others, so that the images can "carry their message within themselves."[20] This is no different with digital images of Mars. Annotations provide a kind of visual parsing that directs a community of observers to focus on a particular aspect of an image and determine salient features. They can also be used to inscribe these visual coding practices into the image itself. Annotations thus function as a *drawing as* practice that directs collective attention and embeds a theory into the image, encouraging others to *see* that object *as* the same kind of thing that the author suggests, whether a laminating unit or a related stratigraphic section. In successive iterations, an interpretation is developed as and through the way it is *written onto the image*. Writing on images, drawing on them, drawing out relevant elements of the scene, inscribes an analysis onto the alien landscape. In doing so, it transforms these images into maps.

Scientists on the Rover mission frequently use the annotation tools in Microsoft PowerPoint to identify and draw teammates' attention to potential geological structures like contacts, units, outcrops, or bedrock as they encounter them. But such interpretations are not always limited to a single image at hand; they can also tie images together as depicting the same type of thing. For example, several months into the exploration of Victoria Crater, Ross presented a colorful decorrelation stretch at the weekly End of Sol science teleconference meeting, using Greek letters to direct the team's attention to what he identified as distinct layers on the crater rim: "I was just looking through some of the recent color images we have of Cape Saint Vincent and noticed something interesting. . . . I labeled them the alpha, beta, and gamma layers. . . . I wondered if anyone else has noticed this?"[21]

Annotating units at Victoria Crater with Greek letters adds two layers of interpretation. It certainly distinguishes the units from each other and from other features visible in the image. But these letters also correspond to three similarly labeled colored units found at Endurance Crater, which *Opportunity* examined before arriving at Victoria (fig. 4.3). Ross thus visually demonstrates through cross-referenced annotations and talk that these units are the same type of thing. He simultaneously *draws* Victoria Crater *as* composed of distinct units, even as he *draws* Victoria Crater *as* Endurance Crater. He thus sets up a very particular

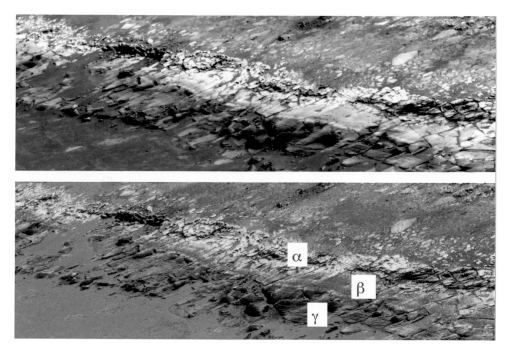

Figure 4.3. Slide depicting false color and decorrelation stretch images of crater ledge labeled with Greek letters to distinguish between geological units. End of Sol presentation, April 4, 2007. Used with permission of Bill Farrand.

seeing as experience at Victoria to support a coordinated analysis of the crater's layered units.

Such annotations and visual identifications are frequently taken up as symbols or even as object proxies to direct continued engagements. For example, when searching for a piece of ejecta accessible to the rover on the rim of Victoria Crater, Ben spoke up on the SOWG line, requesting a Pancam image of "something big and dark purple." He also requested particular filters that would allow him to locate such material, because "what we wanna use this [image] for is to just be able to pick out the dark purple and reds."[22] Nothing on Mars is actually purple to the human eye. What Ben is referring to is objects that display in purple when he uses a conventionalized false color algorithm, because those objects possess a property he wants to examine further. The result of this search was the rover's approach to the rock target Cercedilla and Ben's analytical work discussed in chapter 3. Similarly, when Ross labeled crater layers alpha, beta, and gamma, those names and their false colors were used in conversation to plot the use of the RAT instrument and spectrometer suite on descent into Victoria Crater.

As a particular color comes to be associated with a particular geological aspect such as geochemical properties or a stratigraphic section, new objects begin to be referred to by their representational forms. This metonymy here links a representational convention to its object, so that the annotation or other identification becomes not just the marker but also the inherent characteristic of a class of objects. Such a classification can move individual observations from the status of a singular view to that of a salient phenomenon featured on the evolving geological map. Recalling Susan's decorrelation stretches of Tyrone, Wishing Well, and Arad, such conventions may more easily draw together distinct and even distant observations, now seen as the same types of things, into a narrative whole. The culmination of this work, the geological map, is a kind of drawing work that enables Rover scientists to piece their local observations—whether Pancam multispectral work, Navcam images, or MiniTES stares—together into a regional vision consistent with an underlying geological "hypothesis." As the relevant classes of objects develop through these iterative observations, these ways of seeing impress themselves on the field. Once drawn this way, they cannot be unseen.

Maps for Interacting: "We Can Never Do a Drive without an Image"

Scientists' annotations and geological maps write interpretations of the landscape onto images, turning them into maps of the region. But these maps not only represent a scientific interpretation of Mars, they also have direct implications for interaction with the surface, whether as maps showing where (or where not) to drive, geological maps indicating which question marks to characterize next, or screenshots depicting instrumental targets. Drawing onto images, or drawing out features from them for attention, also helps to plan the rover's future path at each stop where images are taken. Such parsed visions of Martian terrain can present paths for movement, points for interaction on Mars, or simply blank spaces in the map to fill in. As one of the team members explained to me, "We can never do a drive without an image."

For example, several weeks before *Opportunity* reached Victoria Crater, the members of the science team hosted a discussion at their weekly End of Sol meeting about the "campaign" or suite of observations they would do at the crater to address their questions. Stewart, a prominent geomorphologist, kicked off the discussion:

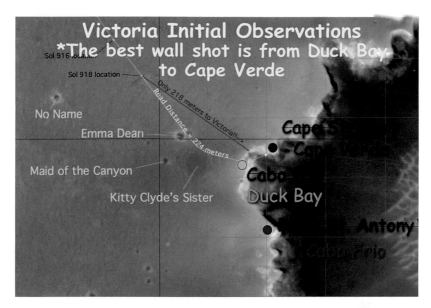

Figure 4.4. Red, green, and blue dots on an orbital image to decide on drive direction. *Opportunity* Long Term Planning report, September 22, 2006. Used with permission.

Basically we need to mostly focus today on choosing what will be the first pan[orama] position, pick a promontory close to [where the rover will arrive at Victoria]. . . . So then further down the road we can debate how many stops to do along the way to hit some promontories [with Pancam images] that will help convert the crater into a three-dimensional mapping of the stratigraphy, which we really haven't been able to do to date. So that will be a brand new thing for the mission, is that you've got so much continuous outcrop that you can actually look to see the lateral changes in the distribution of [features].[23]

The goal of the meeting was to discuss where and how to take Pancam images to answer stratigraphic questions. *Opportunity* will "hit some promontories" where it will take Pancam images to see the outcrops and to create a three-dimensional map, both of which will assist in identifying strata. To ground this conversation and help determine where and how those images should be taken, Stewart deployed a visual aid: an orbital image of the region by the Mars Orbiter Camera (fig. 4.4).[24] Like Joseph, he also drew on the orbital map, using annotations to show others what to see. But unlike Joseph's map, which synthesized existing scientific observations of the region, Stewart's annotations presented possible interactions with the surface in terms of driving and imaging: "The next slide then, we see a map of the crater with two obvious ways to go. The yellow route is longer, and I would advocate it just simply because in the northeastern part of the crater just based on the MOC data, we see the most . . . layers of strata."[25]

Stewart here uses colored lines not to show mineralogy, but to indicate possible rover drive routes that, to his eyes at least, are "obvious ways to go." The question of where to go and where it is best to arrive on the crater's rim is particularly important for maximizing Pancam imaging opportunities. Stewart outlines the possibilities through annotations as well, using red, green, and blue dots and talking through the various options in terms of what could be visible at each location:

> What part of the crater do we hit first? In the three colored spots you can imagine going to one of those three positions. The advantage of the green dot is that you get an excellent view of both the wall next to the blue dot as well as the look toward the red dot. . . . The advantage of the blue is that we can look back toward the red. . . . I would suggest that if you got to the red dot, you're looking pretty far away. . . . That view is not significantly improved by moving toward the blue dot. . . . So if we imagine that the first set of images would be the crater as well as the adjacent cliffs, it may be best to go to the green dot.[26]

Debating where to go, other scientists on the line spoke up. One suggested that the green dot would be a good place to do the "glory panorama" described in chapter 1. Another hoped the Pancams could view the striations on the opposite side of the crater, but Stewart countered, "The problem is that long gazes across the crater, while somewhat spectacular for the [popular] press, may not be very helpful to us." The far wall would be too far away for the image to be "helpful" to "view the cross-bedding." Finally another scientist spoke up to suggest that "the view from any spot, that first view, is going to be more informative than the orbital imagery."[27]

This kind of activity is common when deciding on targets for interaction. Because Maestro, the rovers' activity planning software, did not originally permit sharing or storing target locations, team members turned to tools at hand such as PowerPoint and Photoshop to place arrows, circles, or colored dots onto images to mark their preferred target locations. They then circulated a screenshot among the team to project and record the placement of an instrument or drive path. In fact, instrument operators regularly demand screenshots from sci-

Figure 4.5. Red dots on a Hazcam image indicate MiniTES stares of objects at those locations. SOWG meeting, May 16, 2007. Image credit: NASA/JPL/Arizona State University.

entists during planning meetings so that they know where to place the requested observations. Placing red dots, instrument targets, or other annotations onto an image projects an interaction with the landscape, then later comes to record where those interactions took place (fig. 4.5).[28]

Naming also plays a role in *drawing* Mars *as* a map. Whether mapping the Americas or planetary bodies, naming new territories is at the same time a cartographic and a political venture.[29] But while official names for Martian sites must be submitted to and approved by the International Astronomical Union, here I note a quotidian sense of naming on the mission, as local targets for rover interactions such as nearby rocks or patches of soil are given colloquial names to distinguish them from each other and to coordinate robotic activities. They conform to team-specific naming schemes that correspond to each new area or region the rover explores. For example, since it was initially approached near Saint Patrick's Day, Tyrone is named for a county in Ireland. Home Plate and several other targets are named for baseball, many team members' favorite sport.[30] Selecting targets during African American History Month or Women's History Month has resulted in naming some of the most important rocks, cobbles, and other features on the mission after players from the historical women's and African American baseball leagues, such as Gertrude Weise or Fuzzy Smith. Places around the rim of Victoria Crater like Cape Faraday are named for places, people, and ships encountered during Ferdinand Magellan's voyage of discovery, while Mitcheltree Ridge was named for a JPL colleague who died in a car accident.

A few of these targets may feature in publications, but these names are not otherwise meant to travel. Referring to these names among team members evokes a familiar sense of shared place and experience on Mars. Names thus formulate membership even as they serve as a spatial and temporal mnemonic device. Importantly, names are given to targets: those features that the rover will investigate further through imaging, instrumental techniques, or driving. As such, naming is one of the visual practices of annotation that records and projects robotic interactions onto the Martian landscape.

The naming convention was established in the early days of the mission by social science researchers, who noted the need to distinguish between different kinds of robotic activities and between particular targets for the rovers' actions, both in the software and in scientists' speech.[31] They described this problem as one of developing local ontologies, identifying objects on the Martian surface that would be the targets of different types of robotic interactions, and developing a shared vocabulary to help scientists and engineers communicate effectively. However, target names also serve to *draw* Mars *as* a map. Names ground and document both the team's interactions, in terms of coming to agreement on a target location, and the rover's interactions, in terms of performing the requested observations on Mars.[32] Names additionally record and project scientific interpretations based on these observations. Like Ben's "dark purple" rocks or Ross's alpha, beta, and gamma labels, individually named targets like Wishbone, Tyrone, and Rogan may become elevated to stand for an evolving "class" of materials on Mars.[33] Writing target names onto images therefore locates potential options for continued interaction. In this way, target names, inscribed among other annotations such as dots and lines, identify not only where the rover is and what is around it, but also what it might do next.

Thus visual annotation and target naming bring other team members into a shared members' vision of the landscape, as prepared for a particular kind of interaction. The images that circulate with red dots and target names are critical for communication between the scientists who design the experiment and the engineers who implement it, since instrument operators use these marked-up images to point their instruments. Rover team members refer to the coordination of "red dots" and "screenshots" to be sure that the scientist who requests an observation and the instrument operator who will deliver it are both "on the same page." This annotating image work is so central to rover interactions that failure to implement an observation correctly is often ascribed to visual miscommunication. For example, the team spent considerable time in a SOWG meeting determining exactly which one of a variety of nearby silica-rich cobbles it would

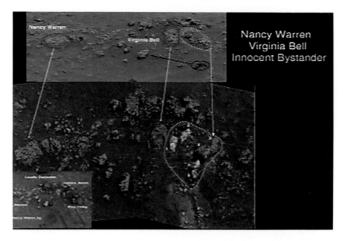

Figure 4.6. Postcrush images and annotations across multiple images of high-silica cobble targets named Nancy Warren, Virginia Bell, and Innocent Bystander (circled in green). Annotations both name and provide correlations between true color context images, false color close-up high-resolution images, and MiniTES spectral data overlaid on a Navcam image (inset at bottom left). Targets were named after players on the American Women's Baseball League team the Kenosha Comets—and after the circumstantial failure to crush the correct cobble as planned. This image was circulated in an End of Sol Long Term Planning report on June 27, 2007. Image credits: NASA/JPL/Cornell/Arizona State University. Used with permission of Steve Ruff.

be best to crush (fig. 4.6). But owing to the difficulty of planning such a precise maneuver with a five-wheeled robot at a distance of millions of miles, *Spirit* ended up crushing an adjacent rock by mistake, which the team retrospectively named Innocent Bystander. At the next day's SOWG, the Chair explained to me, "I would maintain that the reason we didn't crush [the target] is because we didn't have a good idea of where we were. . . . We couldn't visualize it."[34] When a different observation of a rock target failed because no screenshot was precirculated, I witnessed the SOWG Chair wonder aloud, "Why was I so confused [about that observation]?" When I asked if it was perhaps because there was no visual to accompany it, he vigorously agreed. "If there were a visual it would have been completely obvious,"[35] he said: "completely obvious" because the group could have *drawn* Mars *as* a map for rover interaction.

Computational Coordination, Robotic Interaction

Rover maps are not made with dots, lines, and names alone: they may also involve digital image processing. Over nine Earth months, the Rover team commanded

Opportunity to drive up to each ledge around Victoria Crater in turn, and from there to snap high-resolution Pancam images. In addition to being breathtaking vistas that were also useful for analyzing crater stratigraphy, these images had another intended use, as Stewart described above: to create a three-dimensional model of the crater. One of the central purposes of this model was to inform the decision of how and where *Opportunity* should try to enter Victoria Crater. In the Geographic Information Systems (GIS) laboratory at a large public university in the United States, Mars Rover Participating Scientist Tom's graduate students and staff spent much of the summer of 2007 analyzing hundreds of Pancam images for this task. In their work, the techniques of coordinating observations and *drawing* Mars *as* a map for rover interaction are achieved not through field experience, as in Joseph's geological maps, but through computation.

Geomorphologists frequently rely on the Pancam's stereo capability—the fact that the rovers have two eyes—by transforming an image into a three-dimensional anaglyph and donning red/blue glasses to identify stratigraphic sections. But those interested in regional topography go a step further, using the parallax between images taken by the right and left Pancam eyes to generate a three-dimensional "mesh" of the terrain. When characterizing an object as large as Victoria Crater with such distant promontories, the thirty-centimeter displacement between these two cameras is not enough. So the team drives the rover several feet between shots, taking pictures from two displaced locations to produce what they call "long baseline stereo" imaging. This essentially translates to stereo images taken from a wider stance—or baseline—than the distance between the rover's eyes. As Tom explained it to me, "Long baseline stereo is very important for this mission, because our rover only has thirty centimeters of . . . base [between the Pancam's eyes] . . . but [to analyze the crater] the base is too small. You can't make a rover that wide! You have a rover drive five meters here and five meters there, you can get a longer base. When you look at two pictures with wider angle you get a higher degree of accuracy."[36]

Over the several months at Victoria, then, SOWG meeting participants carefully allocated enough time and bits for *Opportunity* to take several pairs of high-resolution Pancam images of the crater from each of the promontories, driving five meters between each set of photographs (fig. 4.7). The resulting images are then numerically analyzed to generate a three-dimensional sense of the terrain. By selecting common points shared between the two images—for example, a rock visible in both images—and comparing the differences between these "tie points" owing to the parallax caused by stereo vision, computers can calculate the depth of the scene and from there generate a topographical model

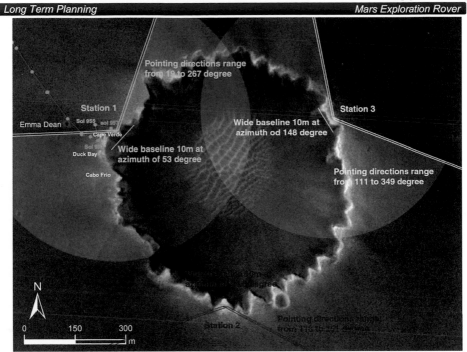

Wide Baseline Mapping Option 1

MER-B
Opportunity

Long Term Planning

Mars Exploration Rover

Pointing directions range
from 19 to 267 degree

Station 1
Emma Dean   Sol 955   sol 957
Cape Verde

Station 3

Wide baseline 10m at
azimuth od 148 degree

Duck Bay   Wide baseline 10m at
              azimuth of 53 degree

Cabo Frio

Pointing directions range
from 111 to 349 degree

N

0   150   300
              m

Station 2   Pointing directions range
               from 113 to 351 degree

Figure 4.7. One of the proposed plans for acquiring images toward long baseline stereo mapping of Victoria Crater. This proposed design was not implemented. End of Sol presentation, October 11, 2006. Image credit: Mapping/GIS lab, Ohio State University.

of the surface of Mars. The result is a "terrain mesh" or a "digital elevation map" (DEM)—a three-dimensional model of the surface of Mars (fig. 4.8).[37]

Long baseline stereo image analysis takes a considerable amount of work and specialist vision. In Tom's GIS lab, about ten graduate students are constantly hard at work identifying tie points between images, clicking on matching rocks across stereo images and coloring them in by hand to identify them to the computer as the same rock. One student, Ying, described her process this way: "I look at an image and judge whether it's the same."[38] Her colleague Yao described his project like this: "First, generate anaglyphs [stereo], use experience to find identical rock . . . if you look at the same thing for many, many times you will see the same thing."[39] This language again is reminiscent of much of what we have already heard from Ben, from Susan, or from Pancam calibrators: they all characterize their work with digital im-

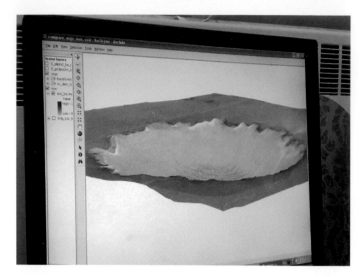

Figure 4.8. Three-dimensional terrain mesh of Victoria Crater, composed through Pancam image processing. Author's photo.

ages as requiring looking, judging, using "experience," and looking at the same thing "many, many times." It also recalls the classifying work of identifying commonalities, this time not across a region to generate a geological map, but across image frames.

The Pancam long baseline stereo observations require much in the way of human-image interaction and digital labor in order to produce stereo views, but other engineers and scientists use software to automatically locate tie points between stereo images (fig. 4.9). Such tie points serve to identify those observations that can be coordinated, by verifying that a feature in one image is the same as a feature in another. Although this coordination process is digitally achieved, the steps in this process can still be cumbersome in terms of manpower. Sometimes the software identifies tie points incorrectly, requiring manual correction and cleaning of the image. For example, it is very difficult to teach the software to tell the difference between the Martian sky and the Martian ground. One image processor informed me that often "the software will find a tie point in the sky," but when trying to correct this with some kind of optimization program, he realized there was no predictable way to identify the horizon, since it was not always straight or even a regular color. A computer scientist on the mission was working on precisely this problem and demonstrated for me his own program for producing three-dimensional views that began with the systematic removal of the Martian sky from his input image data (fig. 4.10). As with the examples

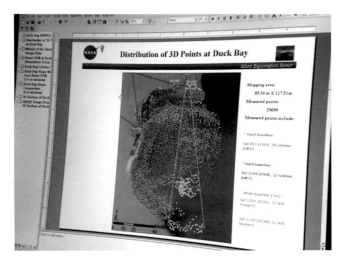

Figure 4.9. The "tie points" generated from the Duck Bay wide baseline stereo Pancam imaging campaign. Dots indicate points that were positively identified as correlated across Pancam images taken in stereo at three different positions. Author's photo.

Figure 4.10. Removing the Martian sky in preparation for three-dimensioinal processing. Image credit: NASA Ames Research Center/NASA/JPL/Caltech. Used with permission.

from the scientific side of the mission, *drawing as* is just as much a question of circumscribing which features are salient and of interest as of drawing certain features and details out of the picture altogether.[40] In this case the judgment of salience affects the computational coordination of images across different observations and files.

Tom's laboratory work with images of Duck Bay reveals yet another way of disambiguating rover image data more common to the operations side of the mission. The aspect that must be acquired and transmitted here with its various constraints and possibilities is one that reveals not spectral or morphological properties of Martian rocks and soil, but rather the topography of the region so as to determine where and how the rover can drive. This labor-intensive digital work is essential to *drawing* Mars *as* trafficable terrain. As Yao explained it, "They cannot let the rover go somewhere with no measured points!" Using the Pancam-derived DEM slope data, then, Tom's colleague Li prepared a color-coded slope map to present to the team (fig. 4.11). He described it to me as "the contour map to show people the [slope]. . . . This red color and this orange color it is not safe to drive."[41] Bo, one of Tom's graduate students in the lab, followed up: "This slope map will be very helpful for these operations guys." I visited Tom's lab the very day before the decision about how and where to drive into the crater was planned for the morning SOWG meeting, so the lab was buzzing to get all the most recently acquired images processed in time. "Tomorrow at 9:00 [a.m.] it's gonna be useful," said Bo; "otherwise it's not gonna be used."[42]

Tom, Li, and Bo's vision of the Martian surface, including the transformations of rover image data that they will circulate for the rest of the team to see, *draws* Mars *as* topographical terrain primed for rover interaction. They used the DEM data that they produced to generate a slope map, which proved central the next day in deciding where and how *Opportunity* could safely descend into Victoria Crater. At JPL, too, Rover Planners frequently use digital elevation data to produce a virtual reality simulation, in which operators have a sense not only of what it looks like around their rover but also, importantly, the undulations of the terrain. Jesse, a camera operator at JPL, echoed Yao's explanation with, "The rover doesn't move anywhere without taking stereo data and processing stereo data."[43] The software that enables rover driving imports this image-derived digital elevation data as a terrain mesh and overlays Navcam images on top to create a virtual environment for drive planning (fig. 4.12). Versions of this software exist for those who operate the Pancams or other instruments; using DEM data, the computer can instantly color a patch of a Navcam image to show where the

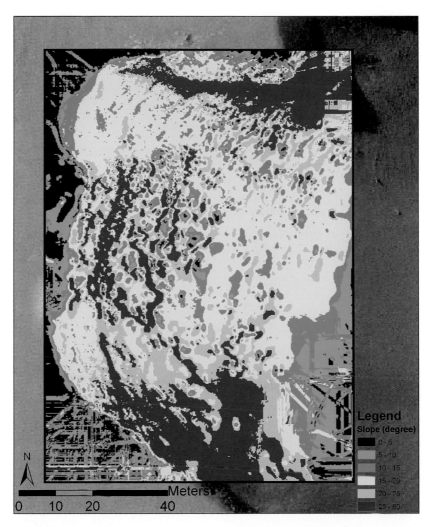

Figure 4.11. Drawing Mars as trafficable for the rover. The slope map for Victoria Crater ingress: the result of the wide baseline stereo mapping Pancam campaign. Image credit: Mapping/GIS lab, Ohio State University.

rover can reach in order to place an instrument (fig. 4.13) or place a colored block over the Martian terrain to show where a Pancam image will eventually be taken (fig. 4.14). Although I was restricted from witnessing firsthand how users interact with this software, team members report that they use these tools regularly and rely on them daily to model how and where the rover can drive or place an instrument.

Figure 4.12. Using digital elevation data as a terrain mesh, rover software overlays Pancam or Navcam images and the robot's position to animate a three-dimensional sense of the rover's location. Maki et al., "Operation and Performance of the Mars Exploration Rover Imaging Service on the Martian Surface." *Opportunity* Press Release, January 17, 2004. Courtesy of NASA/JPL.

Figure 4.13. Hazcam image colored in to show where the rover can reach. Colored dots indicate targets. Author's photo.

Figure 4.14. *Opportunity* Navcam mosaic with blocks indicating where Pancam images will be taken. Author's photo.

The engineers also frequently combine DEM data, parsed stereo anaglyphs, and their own professional vision to create what are colloquially called "lily pad maps." These images are created digitally and physically by drawing on existing images to create annotated representations of the terrain showing where it is safe for the rover to drive and where it is optimal to soak up solar energy (fig. 4.15). Coloring a region in green and coloring hazardous or poorly lit areas in red, the rover is said to "hop" from green patch to green patch like a frog in a lily pond. Lily pad maps are also used to show where the slope of the terrain faces the sun, good spots to stop in order to accumulate solar power. This technique has proved so pervasive that the rover science activity planning software regularly produces lily pad visions of the surface that are captured in screenshots and circulated among team members when planning a maneuver.

Notably, these images and image-processing techniques not only depict a slope map, scientifically speaking; they depict a slope map that identifies only information relevant to the rover's operation. Li's map uses red, yellow, and green to denote local topography, but the choice of colors represents only what is safe or unsafe for *Opportunity*. Thus *drawing* Mars *as* a topographical map is a question of knowing how the rover moves, navigates, and interacts with the terrain in order to inscribe these images with the point of view, possibilities, and limitations of the robotic body. That is, Mars is *drawn as* tangible and interactable for the rover: what the team calls rover trafficability. I will return to this point in more detail in chapter 6.

Figure 4.15. Drawing Mars as trafficable for the rover, this lily pad map uses Hazcam images of the region near Home Plate, showing trafficable (green) and untrafficable (red) areas. Author's photo.

Conclusion: Interaction Points

Through representational practices, team members transform rover images into local maps suited for a variety of purposes. Joseph draws on top of orbital images of Mars to develop hypotheses about a region's geological history, filling in the blanks for areas the rover has not yet visited. Stewart places red, green, and blue dots on an image to decide where the rover should drive. Tom's graduate students painstakingly identify tie points across hundreds of Pancam images to develop a slope map for *Opportunity*'s crater ingress. As scientists and engineers across the mission trade images with dots, lines, names, and colors drawn on them, they not only bring disparate observations together in a single, interpreted visual frame, they develop shared visions of what the rover is currently confronting and what it should do next.

There is a direct relationship between how these images are parsed and represented and subsequent decisions for rover driving and observations. The slope and DEM maps of Victoria Crater generated from Pancam images in the GIS lab were indeed used in the SOWG meeting the next day, when the SOWG Chair credited them as instrumental in "nailing down the slopes and the ingress routes" into Victoria Crater. On examining the maps, the team members changed their initial opinion about where and how to drive *Opportunity* into the crater and immediately began planning to implement the drive. Lily pad, Hazcam, and anaglyph images are used to assess obstacles, slope, and drive direction. Even the question marks on Joseph's geological maps and the dots on Stewart's maps direct attention and invite investigation. Such work with images affects how objects in the rovers' visual fields are talked about, interacted with, and moved about in. All rover interventions are predicated on such representations. Maps work.

But if maps work, is it because they are produced through considerable visual intervention on the part of the scientists and engineers who develop them. That is, rover interactions on Mars are predicated on techniques of purposeful image construal on Earth. Instead of deploying computational techniques to reveal different aspects or make categorical distinctions within an image frame, these techniques of annotation and computation bring such aspects together to draw Mars as tangible, interactionable, and knowable. These representations are certainly predicated on object interventions, as Ian Hacking would suggest.[44] After all, the rover must first take an image, an instrumental reading, or conduct a drive to acquire the image. But we must equally note that these object interventions are predicated on representational interventions as well. It is the techniques of *drawing as* that not only produce new aspects for *seeing as*, but also produce possibilities for interaction. Representing and intervening are iterative activities—and practical, material, interactional ones at that.

Collective Visions

Assembled on a teleconference line for their weekly End of Sol meeting, the Rover scientists turn to a picture of *Spirit*'s location taken by HiRISE, the high-resolution camera in orbit on the Mars Reconnaissance Orbiter. The scientists are trying to agree on how to drive *Spirit* to the southern edge of Home Plate as quickly as possible so they can, from there, move on to a location where it will be safe to spend the next winter and from where they can access an intriguing site to the south, an area they call Von Braun. A scientist who is a member of *Spirit*'s Long Term Planning group and a geochemist by training opens the conversation by directing his colleagues to an orbital image of Home Plate, drawn on using arrows, circles, and lines generated in PowerPoint to demonstrate a multistage approach to exploring the region (fig. 5.1A–C). This scientist hopes that looking at the images together will "mak[e] sure we have agreement with the LTP and SOWG Chairs" and allow them to point out the locations "where there might be some controversy" about how best to get there. "I hope there's no controversy," replies another LTP Lead, a geomorphologist, pointedly. "We really want to get to the southwest corner of Home Plate as soon as possible."

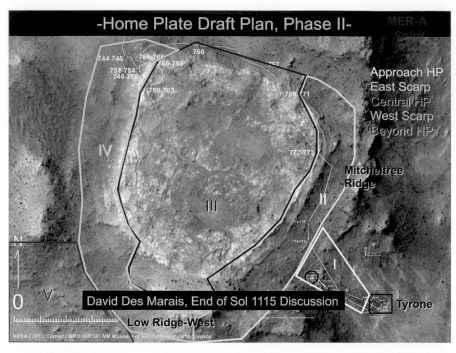

-Home Plate Draft Plan, Phase II-

MER-A
Spirit

744-745   764-765   766
751-754   755-758
746-750
759-763   767
768-771

IV

III

772-773

Approach HP
East Scarp
Central HP
West Scarp
Beyond HP

Mitcheltree
Ridge

774-775

779-778

N

0   V

I

David Des Marais, End of Sol 1115 Discussion

Tyrone

Low Ridge-West

NASA / JPL / Cornell / MRO-HiRISE / NM Museum of Natural History and Science

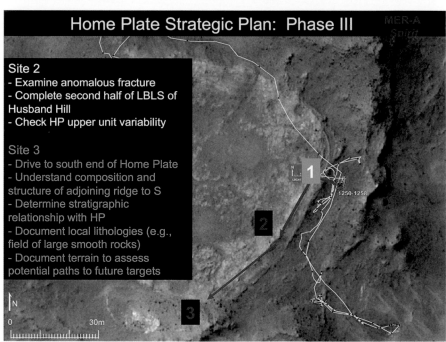

Home Plate Strategic Plan: Phase III

MER-A
Spirit

Site 2
- Examine anomalous fracture
- Complete second half of LBLS of
Husband Hill
- Check HP upper unit variability

Site 3
- Drive to south end of Home Plate
- Understand composition and
structure of adjoining ridge to S
- Determine stratigraphic
relationship with HP
- Document local lithologies (e.g.,
field of large smooth rocks)
- Document terrain to assess
potential paths to future targets

N

1

1250-1258

2

N

0   30m

3

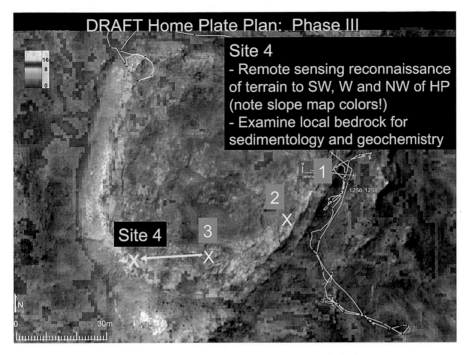

DRAFT Home Plate Plan: Phase III

Site 4
- Remote sensing reconnaissance
of terrain to SW, W and NW of HP
(note slope map colors!)
- Examine local bedrock for
sedimentology and geochemistry

Figures 5.1A–C Three iterations of the Home Plate planning map, End of Sol presentations, June 27, July 18, and September 12, 2007. Base image credit: NASA/JPL/University of Arizona. Courtesy of David J. Des Marais.

On outlining these dual goals of a strategy both for *Spirit* and for the present discussion, the geochemist displays "a chart that tries to carefully map out the drive times" and marks on the orbital image how long it will take to get from point to point to arrive at their goal location. The drive times will be important for elaborating "what would be a reasonable set of science objectives that could be accomplished reasonably within a twenty-sol block" around the necessary drive sols. Identifying what is "reasonable" involves balancing scientific goals with operational constraints such as how long it will take *Spirit* to get somewhere suitable for winter survival and the robot's capacity to manage slopes and soils with its broken wheel. Yet another scientist on the line confirms that the Rover Planners are engaged in visual analysis of the orbital and rover imagery to "look at the evidence and see what we're up against" In terms of generating driving projections; the engineers will report back to the scientists as soon as this task is complete.

Turning to what science they want to do along the way, team members recommend Pancam images, MiniTES observations, and APXS measurements.
One scientist advocates an approach that is less pressed for time as "a defensible
objective from the point of view of field geology," and another wants to know
"What happens if we get there sooner or get there later?" At the end of the conversation, having compiled a list of science requests that could be accomplished
during the drive, that avoided controversial encounters, and that are therefore
considered "reasonable," another LTP Lead—Roger, an astrobiologist—refers
to the orbital image on-screen and tells his colleagues, "I would just suggest that
we annotate this diagram in some way . . . to capture what you're saying."[1] The
result of this discussion of images is more images: drawn on, marked up, colored
in, then presented at SOWG meetings in the routine LTP report and circulated
among the team members.

The previous chapter described how mapping and other *drawing as* practices
construct visions of the planet Mars primed for robotic interaction. However,
much work in critical geography has also brought our attention to how aspects
of power and social relations are enlisted in the mapmaking enterprise. Whether
mapping planetary bodies or colonial boundaries, the representational choices
inherent to mapmaking reveal not only underlying theoretical commitments,
but also resource distribution and networks of authority.[2] Such ordered visions
are often imposed from a centralized authority, such as a state power. But examining this representational work backstage on the Rover mission reveals visualizations produced from the bottom up, consistent with the mission's local form
of social order.[3]

In this chapter I will focus on how such image work—making maps, placing
targets, and annotating with hypotheses or drive opportunities—is enrolled in
the production of team solidarity and collectivist dynamics. That is, images that
coordinate drive planning, image acquisition, and even scientific interpretations
do the work not only of representing Mars primed for interaction, but also of
managing the team. While I recall much of the material from previous chapters
here, my emphasis is on how visual planning is part of establishing a collective
vision: producing and then transgressing, minimizing, or building bridges across
disciplinary distinctions and reproducing the team's collectivist orientation.

The examples below will examine various moments of disagreement or
differences of opinion on the mission, with an eye to how those conflicts are
resolved. It is therefore important to note that just because a group is consensus
oriented does not mean it is averse to disagreement. As anyone involved in a
consensus group knows, consensus requires disagreement, even invites it. Stud-

ies of collectivist-based social movement groups have shown that the balance of inviting all opinions, then narrowing down to one of them, is a complex process with a variety of means of handling conflicting views.[4] As I described in chapter 1, the Rover mission has confronted these problems by developing roles, rituals, and problem-solving strategies that situate moments of conflict within the team as a constructive part of the process and not as a divisive experience. It is not so much that the group is conflict averse as that it has developed very particular ordered ways of managing conflict, with their own internal logics. The cases I describe below represent either the locally ordered approach to disagreement or moments of breach in which that same social order becomes visible through its absence or direct invocation.

Depicting Consensus: Long Term Planning Maps

In chapter 4 I described how image annotations transform images of Mars into maps that may construct geological narratives or suggest future interactions with Mars. Such different *drawing as* practices produce different visions of the terrain, prepared for different kinds of interactions and interpretations. But another form of mapping on the mission reveals an additional analytical layer to mapping: that of anchoring and recording consensus. Long Term Planning maps are produced in the strategic discussions hosted at End of Sol meetings; their annotations do not naturalize a vision of the terrain for investigation so much as they record and naturalize the continuing social achievement of deciding on the rover's activities over the coming weeks. As activities, targets, and object identities are written onto an image and circulated, presented as arising naturally from the terrain, these images help to craft and sustain a shared vision of Mars exploration among MER team members.

Long Term Planning Leads regularly insert orbital or ground-based images into their End of Sol presentations to enhance their conversations about strategic goals for each rover. As team members ask each other "Which of these is our next objective? This, this, or that?" they use circles and dots on the image to articulate different possibilities for the rover's engagements. They use this conversation around a suite of images to converge on a plan for the next round of activity (fig. 5.2A–D).[5] Indeed, images are so crucial for these conversations that at one point, when a strategic discussion was under way, an LTP Lead remarked, "It may be more useful for the discussion to keep the map up on the screen. . . . We need to converge towards [sic] some kind of priority."[6] The result of these conversations is more annotated images that capture the convergence in the con-

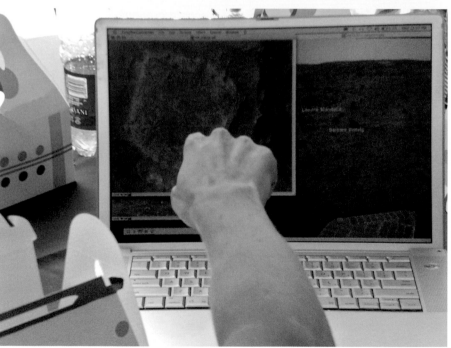

Figure 5.2A–D. A rare in-person planning discussion among LTP Leads, a SOWG Chair, and collaborating scientists. Gathered around a computer screen displaying the orbital image of Home Plate, they coordinate talk, gesture, gaze, and annotation toward the combined activity of visual sense-making and strategic planning. At the Seventh International Mars Conference, Pasadena, CA, July 11, 2007. Author's photos.

versation, using arrows, boxes, circles, or question marks. These images are then imported into LTP lead presentations at the opening of every SOWG meeting to make sure everyone is still on the same page. Circulating thus, they become part of the local political economy of images that reinforces a consensual understanding of Mars and collective decision making about where the rover is, what it is looking at, and what the team has decided to do about it.

An example is the series of images of the Home Plate region that Roger, the LTP Lead throughout much of *Spirit*'s Home Plate campaign, produced and circulated throughout the Earth year 2007, between *Spirit*'s second and third winters on Mars (fig. 5.1A–C). Using a single orbital image taken by the HiRISE orbital camera in October 2006, Roger drew and redrew projected paths, targets, locations, and proposed phases of exploration. Presenting one map at an End of Sol meeting, he titled it "Draft Strategic Plan," noting first that "the emphasis here is on draft" but also that the map already encompassed "a fair number of inputs from a fair number of perspectives." He introduced the orbital image as "the background map, the base map of a lot of what we're going to present here" and "the traverses that *Spirit* is doing now." He then annotated sections of Home Plate as "Phase I," "Phase II," and "Phase III" to capture "an approach to thinking about the exploration of Home Plate, sort of in time sequence." Roger then drew several possible trajectories for the rover: the preferred one, moving "clockwise around Home Plate ending up sort of at six o'clock [position]," and the backup plan, "trying to get around Home Plate going up onto the top in a counterclockwise position."[7]

Already incorporating "a fair number of perspectives" from previous conversations, this image also generated lively discussion. One scientist suggested that *Spirit* "just go around Home Plate, to heck with the top, and just get on with the West side," while others debated the importance of investigating the eastern or northern rim. A slope map was circulated alongside the annotated image to demonstrate "the source of our optimism" that the rover could make it onto the top of Home Plate. Ultimately the scientists agreed to discuss specific objectives in small groups to best inform the observations *Spirit* would need to take at each step.[8]

Annotated maps become the subject for discussion as the team members try to articulate what they should do next, but these maps also reflect agreed-on decisions for activities undertaken by the whole team. After their strategic discussions, LTP Leads will assemble and circulate an updated annotated image that reflects the conversation. After an End of Sol meeting, for example, an LTP

Lead suggested "that we annotate this diagram . . . in some way to capture what you're saying,"[9] while in another case, after an intensive discussion about where to drive, a scientist requested an annotated image, asking the LTP Lead in charge of the discussion, "Can you send out a description of this, just so we're all on the same page?"

These maps are not static: they are updated based on subsequent discussions. An iteration of this same diagram was circulated a few short weeks later, after Susan's presentation at the Team Meeting in February 2007. All scientists present at the meeting agreed it was time to move quickly through the eastern rim area of Home Plate and to aim for getting the rover onto the top of Home Plate within four weeks: this was recorded in annotations on the image. But between Tyrone and the Home Plate rim, the MiniTES spectrometer identified some high-silica nodular rocks that indicated a potential hydrothermal environment. The team therefore initiated an investigation of what they called Silica Valley that took several weeks. The next time Roger circulated a map, it outlined projected drive targets and directions, showing no intention of moving onto the top of Home Plate until after Silica Valley was fully explored. The next version, devised after the July 2007 Team Meeting, captured the team's desire to move away from Silica Valley and up onto Home Plate as quickly as possible.

The case of many iterative images of Home Plate does not present an instance where annotations failed to project careful team planning or where the team failed to stick to its plans. Rather, each iteration of the image captures a consensus moment in the evolving story of the mission. As Roger put it during a meeting, "The approach we usually take, and it's been very fruitful, is that we have a strategic plan, and then as we approach [our target] that plan evolves. . . . As we approach and acquire our [data] it may be that . . . the strategic plan goes out the window."[10]

Roger's maps therefore depict evolving local conversations and the push and pull between tactical and strategic planning as much as they depict Mars. The images are rarely viewed in series but rather replace each other with new iterations every time a new decision is made or a rover drives farther.[11] However, viewing the series with the benefit of hindsight, the analyst is presented with an evolving story of the mission, the crucial features at each moment faced in the terrain, and the dividing decisions that needed to be overcome. As they evolve and change, these images present a trace of a moment when the team members reached consensus in an ongoing conversation about the Martian environment and their interactions with it.

Targeting Agreement

Long Term Planning maps present snapshots of an evolving conversation, but simpler planning documents similarly reflect or encourage a collective vision of the landscape and may be enrolled in producing a consensus moment about which observations to make. One place to observe this process in action is when the group selects targets for rover interaction. Placing a target is sometimes an individual affair: scientists may do this at their leisure in their local versions of the rover software, and target names are either e-mailed to the Keeper of the Plan (KOP) or assigned at SOWG meetings for input into the software. But when targets must be negotiated as a group, the process of identifying which region to target reveals a combination of visual analysis, social convention, and technical action. For example, after Susan's presentation at the January Team Meeting in 2007, *Spirit* returned to Tyrone to do some follow-up analysis on the area, including taking thirteen-filter Pancam and Microscopic Imager images as well as spectral readings of small nodular rocks (called clasts) visible near Tyrone. The team budgeted only a few days for these follow-up observations before moving back toward Home Plate. The Chair opened the SOWG meeting by declaring, "The primary objective is to get MI and Pancam thirteen-filter [images] of Tyrone, because we didn't get those the last time we were there." The subsequent LTP report included an image that the LTP Lead described thus: "Over to the right is the approximate position for the APXS target, Mount Darwin; on the left is a bunch of blue circles surrounding potential targets pointed out by several people. . . . We might do an additional MI observation in the coming plan, and so we'll hear from those people about the desired target."

Even before the team started to place the MI targets, they were flooded with images. The next slide was described as "a visual summary of what we went over" during the opening review, followed by "an overhead view of our traverse" indicating that "we're on the first leg of the four phases of our exploration in the coming field season . . . number two is the return to Home Plate." On the screen was a view from the rover's own cameras, showing its tracks stretched around it and the names of potential targets (fig. 5.3). These images anchored the tactical discussion about the rover's immediate activities within the context of the larger strategic plan. Through this conversation it became clear that given the immediate need to drive away the next day, there was no way for the team to resolve the issue by appealing to further imaging or taking the conversation offline. This was the last opportunity to get a closer look at Tyrone and its environs.

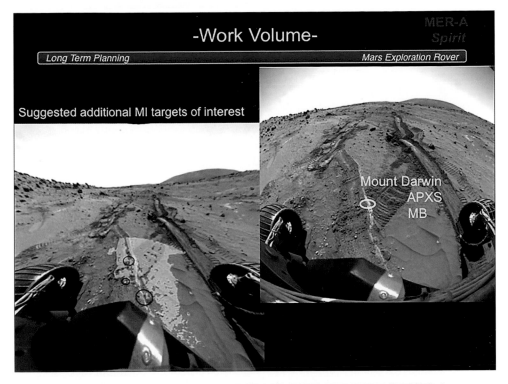

Figure 5.3. Targets and target names placed on a Navcam image to indicate proposed MiniTES observations. Targets outlined in blue are subject to discussion. *Spirit* LTP report, sol 1099, February 5, 2007. Image credit: NASA/JPL/Caltech. Used with permission.

After bookmarking time for the Pancam images, then, the team turned to the image with the blue circles to plan the "culmination observation: the MI of the clast of nodules." When the Chair asked Jane, one of the scientists, to explain the situation and introduce which nodule to image, she said:

> I sent an e-mail with the [annotated images] to [the KOP]. . . . With Susan and Alexa we discussed three potential targets. . . . Susan and Alexa think that maybe target 3 [is best]. . . . I prefer target 1, but I think that in the work volume [the area where the rover can reach] target 1 is the only one reachable; can you confirm that? . . . I would rather have many targets. . . . But Susan thinks that target 3 on that side has the cleanest nodules. . . . The rationale for target 1 is the density and diversity of nodules, target 2 is because of the drift . . . and she thinks it looks cleaner, just visually. I don't think we have any other data.

On the videoconference screen, I watched Susan hand the Chair a piece of paper with an image on it. The two of them and a third colleague from their same institution looked it over, discussing how much dust, if any, was on the nodules. The Chair then addressed the Rover Planners on the line to see if that would resolve the issue:

> *Chair:* Okay, Mark and Rick, do we have any reading on the feasibility of going to any of them?
> *Mark:* Rick is looking at the details. To our eyeballs it looks like either of them should be reachable.
> *Chair:* We have a difference of opinion.

So far, a few local ways of resolving this difference of opinion are evident. First is the appeal to what the rover can or cannot do as a factor limiting which target should be chosen. In her explanation of her preferred target, target 1, Jane notes that the nodule is the only one reachable in the rover's work volume, although she hedges this with "Can you confirm that?" The only way to confirm it is to ask the Rover Planners, whose vision of the Martian terrain is predicated on whether and how the rover can interact with the area, as described in chapter 4. Their assessment could decide one way or the other, but they return with a neutral position: *seeing* Mars *as* trafficable, "either of them should be reachable."

Another scientist spoke up on the line, proposing that perhaps they should see which of the targets could handle a placement from another instrument, the Mössbauer spectrometer. This could result in a coordinated observation between the two instruments, a preferred strategy; it would also invoke the Mössbauer spectrometer's requirements as a limiting factor that could eliminate two of the proposed targets. But the Chair rejected that decision on different grounds. A Mössbauer reading was not a priority for the day: the priorities had been established before the meeting started, as part of the strategic discussion in the End of Sol meeting. To bring in a discussion of an additional spectral reading would transgress the ritual distinction between strategic and tactical discussions: as the Chair put it, "This [the SOWG meeting] is not the place to set the priorities; we have to come into the meeting with one, two, three." So bringing in another instrument's capabilities and limitations would not solve the problem. Another scientist then attempted to appeal to those same set priorities and a chain of previous inscriptions: "I think there is one [target] that is listed as priority number one on at least two of the presentations." This appeal attempted to invoke the visual authority of previously circulated images, but it was unclear whether the previous images included a target.

Ultimately the Chair resorted to making a decision based on feedback or preferences by each of the group members, but then he switched tactics to allow for a consensus to emerge around the target preferred by the most members:

> *Chair:* Who wants number one [target 1]?
> *Ben [on the line]:* Susan and Alexa thought that target 3 should go first.
> *Chair:* So it's kind of two to one in terms of the 'clast mafia.'
> *Jane:* Right. It's fine with me, both have the same information.
> *Chair:* Okay, if we went to number 3 is there any other dissent or discussion needed?
> *Jane:* And you would set up the 1 × 1 [microscopic image] over the big nodule?
> *Chair:* Yes [two-second pause]. Going once, three times, done.[12]

The Chair initially attempted to go through the three targets one at a time and solicit feedback about them, but Ben spoke up on the line with another possible way of resolving the problem. This is a two-to-one split, and Jane has indicated a "preference" but not an ultimatum. If Jane agrees to target 3, they will have consensus. The Chair makes light of the situation by calling the two-to-one split "the clast mafia," using humor to ease any tension. But he thereby offers Jane an opportunity to respond, which she recognizes with "Right." She hedges her initial position, saying "it's fine with me" (indicating, to go with target 3) and "both [targets] have the same information" to indicate capitulation, although she checks that the resulting observation will include the aspect she is interested in (the 1 × 1 texture image of the big nodule). The Chair notes her assent, closes the conversation with the auctioneering phrase "going once, three times," which invites other comments but implies that the tough work of negotiation is over. Once the target location has been decided, then the group can move on to naming it and assembling the code for the observation:

> *KOP:* We need a name [for the target].
> *Chair:* [consulting list of names] How about Pyrenas.
> *KOP:* Okay, I have the target name, we are set on that. . . .
> *Pancam PUL:* Also, if someone could forward me some images of these nodules that we're supposed to be looking at and also an image of the approximate location of the work volume [where the rover can reach], that would be helpful.[13]

Note how, in their attempt to resolve the "difference of opinion," the group made several appeals common among their team. They appealed to what the rover could or couldn't do: the ultimate authority in whether an observation can be planned. Scientists and engineers alike frequently appeal to rover health, safety, and even potential death to resolve differences over what to do, a factor I will return to below. When that was not relevant, they turned to coordinated observations: Which other instrumental observations would this one need to align with, and how might that consideration limit the selection pool? They appealed to the authority of previous discussions, seeking out who, when, where, and why the observation was planned in the first place. These are all attempts to appeal to external factors. They never questioned the authority of the individuals who chose the targets or the scientific validity of their claims, nor was the situation allowed to get personal or tense. Finally, given that none of those appeals were successful, the Chair moved to two options of last resort. He began initiating a fiat by soliciting feedback on each option so as to make his decision, but he switched to a different tactic once Ben spoke up: to see if the two-to-one situation could be amicably resolved by the one scientist's agreeing to the other observation. This avoided the need for a top-down decision and allowed an opportunity for a scientist to capitulate to her colleagues and enable a consensus moment.[14]

Naming and placing a target demonstrates that the team has agreed that this particular area is of interest for further rover work: in this case a Microscopic Image. The resulting image represents a consensus moment in the meeting—a moment when several team members agreed on where and how the rover should interact with Mars. As the target is placed in the rover command software, the moment of consensus around the target is translated into practical action on the surface of Mars: the rover's acquisition of an image. Thus, in order to decide what the rover should do, everyone must see and subscribe to the same plan laid out in the images. As such images evolve in interpretation, then, they become both the mechanism for achieving consensus and the record of that same achievement. Images thus coordinate this work and enable the translation of interactions on Earth into interactions on Mars.

Seeing Distinctions

Both the targeting images and the LTP maps reveal how images are enrolled in the social order of the mission. They are used to anchor strategic and tactical discussions about what the rover should see or do, and they stand as records of teamwide agreement. This latter aspect of image annotations is most evident when team members question their colleagues' images. Although annotations draw distinctions in the Martian terrain, at the same time they may draw distinctions between members of the team that must be negotiated and reconciled within the context of the mission's local order. When disputes arise, then, it becomes clear that the annotations draw more than simply a hypothesis about the terrain.

Long before *Opportunity* arrived at Victoria Crater, Stewart, a geomorphologist, was transfixed by a light-colored ring around the crater's rim, visible in orbital imagery. As the rover neared the crater, he presented an annotated orbital image to his colleagues at an End of Sol meeting, calling their attention to the feature: "Around the whole crater from time to time you can see two discrete ledges, those are labeled just to illustrate it: double ledge, DL for double ledge. . . . On the southwest corner you can see the same feature showing up . . . you can see the area that I've labeled multiple strata."[15]

Geomorphologically speaking, banding around the rim would indicate that the depositional environment was relatively consistent for a particular period. But when Stewart said this, one of his colleagues spoke up to correct him:

> *Scientist:* I would stick with banding, though, as a term. It may not be layering that we're seeing. It may be terracing of some other kind.
> *Stewart:* Okay, well, certainly the doublet seems to be a geomorphologic effect, and it remains to be seen what the other is. But I understand your point, well taken.[16]

At stake here is whether Stewart's annotations represent what the team would agree is a real feature in the environment or simply his interpretation of a feature, an interpretation the team has not yet established. "Banding" is a referentially open term that geologists use to describe merely a visible phenomenon but not its origins, whereas layering or terracing implies some kind of environmental conditions that produce the bands. Stewart capitulated, indicating that what he sees is a real distinction in the landscape but reeling in his nomenclature to something referentially open. Unlike Joseph's maps, which are hedged with his declaration of "sketching" his hypothesis and which incorporate existing observations and interpretations into a meta-analysis of the terrain, there is as yet no agreement about what produced the feature Stewart sees that could justify a bold claim.

Several weeks later, after *Opportunity* had arrived at the crater's edge and begun its imaging campaign, Stewart displayed a newly downlinked high-

resolution black-and-white (single-filter) Pancam image of a cliff face, annotated
with lines to demarcate what he distinguished as different units in the cliff. He
parsed the image verbally for his colleagues to direct their attention to the lines
he traced across it (fig. 5.4):

> These are tall cliffs; we're probably getting ten to twenty meters of exposure
> on the cliff, and in this view of the west face of Cape Verde we can see what
> looks like a massive unit overlain by the breccia of the [crater's] ejecta blanket
> and then underlain by something that looks thin-bedded and quite particulate.
> And then if you stretch [increase contrast on] that image what you can see is
> again that there is a well-defined thin-bedded facies, and one of the questions
> we ask is . . . can you see that, and so far we don't know, we don't have data for
> the outcrop yet.[17]

Using visual and verbal tools, Stewart draws the Martian terrain according
to his hypothesis and presents it to his colleagues. He draws lines to demarcate
different units laid down by different processes through the ages and labels these
strata with their geological names (breccia, thin-bedded facies) to generate a
narrative about the crater's formation. Joseph once called this kind of drawing
"a very useful process because otherwise you have no way of knowing where this
[geological] contact is in this image, but now [once it is annotated] we know
where it is." But simply presenting an annotated image does not mean the image
is closed for interpretation. At the same meeting William, another scientist who
specialized in geomorphology, interrupted Stewart to challenge the annotations:

> The way this slide has always been labeled has been massive above, massive
> below, and then thin-bedded facies. But you gotta keep in mind that this was
> shot from a considerable distance, and the resolution for a lot of our layer-
> ing was really on that scale. So what I really wonder is if what you labeled as
> massive is really massive, and that if we had a vantage point that was closer we
> wouldn't see that there was some finer-scale bedding in that stuff. . . . I think
> that if you look at the images of the other side of Cape Verde . . . I think what
> you'll see is that in fact . . . at more or less the same stratigraphic level of what
> you labeled as massive, there is some layering visible. We got two things work-
> ing against us here: one is the resolution and the other is that it's in shadow.
> And I really question the massive nature of that vis-à-vis what we've seen other
> places. . . . my suspicion is that there's a lot of fine layering that you simply
> can't see.[18]

Figure 5.4. Two R2 Pancam frames stitched together to show the promontory at Victoria Crater (Cape Verde) under discussion. Annotations unavailable. *Opportunity* sols 973, 976, and 977. Image credit: NASA/JPL/Cornell.

William's suggestion is that the image interpretation is premature, since the images are not robust enough to support Stewart's interpretation. But he does not direct his challenge against his colleague's expertise. His counterpoint is full of hedges, using phrases like "my suspicion" and "what I really wonder" to position his interruption as polite, disinterested discussion. He uses expressions like "things working against us" and "this was shot from a considerable distance" to attribute his colleague's potential misinterpretation to Martian conditions and insufficient information. He also blames poor lighting, resolution, distance, and conflicting data for compromising the ability to safely make the visual claims his colleague submits.

To resolve this disagreement, both Stewart and William transitioned the conversation from what the units were to what observations the rover could make that would disambiguate them. The team's geomorphologists and their graduate students planned several suites of images over the coming year, imaging at specific times of day to get the best lighting and shadowing conditions for photogeology, planning Pancam images in the highest resolution possible, and shooting photographs of many cliffs around Victoria Crater to see if these layers were consistently visible around the entire crater. A year later, *Opportunity* drove into Victoria Crater in order to get even closer images of the cliff face and to place the IDD on the rock layers to analyze them more fully. The visual challenge was translated into an opportunity to resolve critical questions through an appeal to more or better observations.[19]

Disciplinary (Di-)Visions

In Stewart and William's case, the disagreement is between two geomorphologists, who share the same professional vision and *drawing as* practices. But the Mars Rover team is composed of a heterogeneous group of scientists and engineers with distinct disciplinary and visual heritages, who must work together to explore Mars. Given the different visual, disciplinary, and other distinctions between members of the team, coordination and agreement constitute an important social achievement. Images, again, serve as a central site for negotiating roles alongside visual meaning: they draw distinctions in the Martian terrain even as they draw distinctions among the various groups that compose the team. Rover planning based on rover images requires an understanding of these social divisions that mediate visual exchange.

The example of target selection given above is instructive for this reason. Jane, Susan, and Alexa come from different perspectives within planetary science, and their different preferences for the targets are based to a large extent on their disciplinary expectations.[20] Geologists are often divided into two camps: mineralogists, who characterize rocks by their mineralogical components and chemical properties, and geomorphologists, who characterize them by visible physical characteristics. The Mars Exploration Rover robots possess a suite of instruments that incorporates both of these epistemic cultures:[21] different kinds of cameras for morphology and different spectrometers for mineralogy. Returning to the target selection example above, Jane's preference for target 1 is based on her interest in the nodular structures, in terms of density and texture: this is a geomorphological issue. Susan's and Alexa's preference for target 3 is based on its "cleanliness"—its relative freedom from dust. This can be important in the spectral analysis these two scientists specialize in, where the dust "pollutes" the spectrum of the object with its own spectral characteristics.[22] Thus the two sets of scientists express different preferences for an image that will allow them to see different things in the target, see different aspects of that "same information" to satisfy their different disciplinary interests and perspectives on these nodules in the context of the Tyrone site. Resolving this conflict to both sides' satisfaction is therefore an important achievement in the context of the mission's collective orientation and local requirement to keep everyone "happy."

This is a classic divide in planetary science, and the team members are constantly attuned to managing this division in practice. Given the choice, they prefer to have it both ways: to satisfy the needs of both sets of scientists. This was the strategy in the selection of the rover's landing sites, which could easily have excluded one or another of these ways of knowing from the equation.[23] Perusal of the proposed landing sites and preliminary scientific investigations of them drew on geomorphological evidence like topography from the Mars Orbiting Laser Altimeter (MOLA) and images from the Mars Orbiter Camera (MOC), as well as spectroscopic evidence like THEMIS and TES datasets. But having two rovers meant the team could select two sites. One, Gusev Crater, was selected for its geomorphological characteristics: from orbit, it looked like a crater with a river running into it, possibly forming a lake. The other, Meridiani, was selected because of the hematite signature that the TES spectrometer detected from orbit. As a mineral primarily formed by interaction with water, hematite was a smoking gun for a mission looking for evidence of past water on Mars. Thus selecting both sites ensured that both communities of practitioners would be satisfied with the mission.[24] Similarly, in the example above, if the MI on target 3 is set up over the largest nodule, then the single image can satisfy both

Figure 5.5. Annotated image presented at SOWG meeting proposing Pancam observations. Used with permission.

sets of scientists, with both ways of seeing. Hence Jane's agreement, with the caveat that the MI be set up directly over the nodule—what she is interested in seeing. Hence also the suggestion to pair the MI with a Mössbauer observation, providing a combination of textural detail with spectral detail that would satisfy both communities' needs.

These different disciplinary preferences can therefore inspire different priorities for observation planning, making it difficult to construct images that appease both sides of the divide. For example, after *Spirit* returned from Tyrone to Home Plate, a SOWG Chair who was also a geomorphologist presented an annotated image (fig. 5.5) in his opening report. A spectroscopist spoke up to dissent:

> *Spectroscopist:* I'm still struggling to understand where we're going and what it is we want to achieve. . . .
> *Geomorphologist (Chair):* Okay, let me recapitulate what we're trying to do. . . . [We are taking more Pancam images to] characterize geometry, cross-bed and textures along that east side of Home Plate. So I would say that we probably have one or two more locations in which to do that, and if we get a really good drive next time maybe it will be one. . . .

Spectroscopist: There was never any discussion of what kind of coverage are we
    trying to fill in. . . . I saw two arrows drawn [on the image], and we arrived
    at a second arrow, and I don't know if we're going to drive farther.

Geomorphologist: The discussion all along was to drive back up to the location
    where the angled cross-bedding is and fill in from there.

Spectroscopist: I think those arrows were drawn pretty haphazardly without
    any discussion of where we are going and what we might be doing.

Geomorphologist: Whether we are at the first arrow or at the top of the second
    arrow, that's not the point. . . . The point is we want to complete an imag-
    ing sequence somewhere between those two arrows.

Spectroscopist: We already have Pancam coverage. . . . How good do we need to
    do this? Why can't we do the imaging from this location and then. . . . be
    done with it? . . .

Geomorphologist: Well, certainly we will be getting images from this location:
    the predrive remote sensing block is certainly supposed to be getting im-
    ages of this section. . . .

Spectroscopist: I at least don't see that . . . . We can see looking back from that
    location on top that [sol] 773 Pancam contains that outcrop we're trying
    to drive to. We've already got [a picture of] it.

Geomorphologist: Yeah, but that's too far away to do [analyze] the geom-
    etries. . . .

Spectroscopist: I guess this is where the minutiae of how much we need to do
    comes in. . . . [B]ut in a tactical reality [of time and bit constraints] we
    can't do Pancam plus MiniTES and get good results.[25]

In this case the discussion centers on differing interpretations of an annotated map displayed in the LTP report at the outset of the SOWG meeting,
which presents an approximate location for an imaging campaign. But the discussion concerns a disagreement over the scientific objectives in the region
around the rover: how well those objectives have already been satisfied and
whether Pancam or MiniTES should be the focus of the day's activities. The
geomorphologist is interested in taking high-resolution Pancam images of this
side of Home Plate to characterize its stratigraphy; the spectroscopist is interested in collecting the spectral signatures of the silica-rich rocks in the same area
using the MiniTES spectrometer. Both are trying to say something about the
depositional environment at Home Plate, but both also worry that one instrument's observations would jeopardize the other's because of the constraints on
rover bits, time, and power. The spectroscopist points out that the rover has

already acquired Pancam images at this location and questions the need to do it again, using up precious time to drive the rover closer to the east side; the geomorphologist counters that the previous images and the predrive imaging are unsatisfactory for seeing the textures owing to distance. In defense of his position, the spectroscopist describes the geomorphologist's concern as "minutiae" and characterizes the "tactical reality" as one that requires prioritizing either Pancam or MiniTES, with no middle ground.

Most interesting about this example is how the conflict centers on annotations on an image: specifically, whether those annotations record a discussion in which consensus was reached over which observations to take. The geomorphologist believes the image does represent what the rover should do based on "the discussion all along." He had annotated the image himself and placed it in the report. But the spectroscopist claims "there was never any discussion of what kind of [Pancam] coverage are we trying to fill in" and says he believes the annotations "were drawn pretty haphazardly without any discussion of where we are going and what we might be doing." The negotiation about this annotated image is not about visual interpretation, then, but about whether the group had agreed on what the rover is about to do on Mars. It therefore reveals the social context of these images' circulation and their valence in the local consensus culture. The image is not technically inaccurate in terms of how it interprets the Martian terrain: indeed, the geomorphologist and his students spent a lot of time composing the map to ensure that the Pancam images were perfectly projected and draped over the orbital image base. The accusation instead is that the image is invalid because it presents an imminent observation that captures what one scientist wants to do but not what all the scientists agreed to do.

The Case of Winter Haven 3

There is no right or wrong answer about where to drive the rover or what images to take, although the team believes that there are better and worse scientific criteria for making decisions about observations and that putting the rovers in physical jeopardy must be avoided at all costs. But most questions do not have a simple yes or no answer. Answering the questions of whether and how *Spirit* should climb onto the top of Home Plate, or how and where *Opportunity* should try to descend into Victoria Crater, depends on how the team processes the images that return from the planet and arrives at a collective decision. On one hand, different kinds of scientists must produce multiple visions of the terrain to inform what the team believes is "the best decision." They do so according to their own *drawing as* practices, consistent with their disciplines. On the other hand, the group must coalesce around shared interpretations of these distinct visions, particularly when multiple images are available. This requires first making visual distinctions between the different mission constituents and then attempting to make a unified decision that elides or effaces those distinctions and can stand as a teamwide decision.

An excellent example is the extended conversation about where *Spirit* should spend its third winter on Mars. After *Spirit*'s second winter, the rover had returned to Tyrone for follow-up observations, then returned to the east side of Home Plate. The plan was to pass the winter on the southern side of Home Plate, where north-facing slopes would guarantee enough power to survive. But with the discovery of Silica Valley en route, as well as the geomorphological structures on the edges of Home Plate, the plan evolved to allow the rover to tarry longer at each location and acquire more observations. There was now no time to make it to the southern edge. The discussion thus shifted toward spending the winter either on a promontory overlooking the southern edge of Home Plate or on the north slope of Home Plate. The South Promontory presented a new vista over what the scientists called "the Promised Land," an area earmarked for exploration to the south of Home Plate, while the north represented a return to a known area, since *Spirit* had spent a previous winter there (and survived). An End of Sol discussion was especially designated by the PI "to get the issues on the table as to the scientific merit" of either site.[26]

The End of Sol meeting started with a presentation by Sarah, a Rover Planner. She showed a Navcam mosaic image (fig. 5.6) with lily pad annotations, drawing Mars as a map for rover trafficability. Speaking for the engineering team, she began:

> The possibility of using [South Promontory] as a winter haven is significantly reduced in my opinion given this set of images. . . . The drive up to the end of this outcrop is full of fairly large rocks, although it looks flat [drags her cursor between red splotches on the image that indicate undrivable areas]. There is a path through there that we think if we constrain things very tightly we could get to the edge of that outcrop. . . . The problem occurs when we hit the outcrop. . . . If you look at the close-up imagery that we have now [displays an image colored in to show various degrees of slope], I can't really find any way that we could park and get any more than twenty-two degrees slope. . . . From all the imagery that I've seen . . . I can't demonstrate with any level of confidence that we actually can reach that parking place for the last segment of that drive.

Figure 5.6. Navcam mosaic view of South Promontory under discussion by scientists and Rover Planners. Spirit sol 1347; annotations unavailable. NASA/JPL/Caltech.

Sarah's conclusion was that twenty-two degrees of slope would not be enough to sustain *Spirit* through the winter, and that while the path to the north-facing area would be difficult driving for the five-wheeled rover, it would be feasible with hard work.

Although Sarah's results were initially discouraging, the PI suggested that the science team consider these findings preliminary and continue with its presentations regarding the scientific rationale for moving south or north. The scientists employed a variety of annotations and visual and verbal parsing to make the case for one or another winter haven site as presenting compelling questions for the rover to answer in situ. As one of many examples, Joseph presented an iteration of his "regional overview geo-sketch map" to make a case for a move to the South Promontory region of Home Plate. Joseph used the same images as Sarah—the orbital HiRISE image and figure 5.6—but his annotations identified the "basic stratigraphy" of the region, including the location and characteristics of the units the rover had already examined and where those could be identified in current imagery (fig. 5.7). Moving from what was known about the region to what was not known, he then identified unusual layers and bedding directions in the few available images of the south (fig. 5.8) and, pointing to the question marks on his geological map, he asked, "What is that ridge on the south edge of Home Plate?"

Figure 5.7. Navcam mosaic of South Promontory, with annotations indicating geological questions (not driving limitations). Used with permission.

My interpretation, and this is the term "interpretation"—that's what basically you go to places to look at things to see if your interpretation is correct or not—is that the top of that ridge is in some parts covered in bits and snatches of the upper unit of Home Plate. If so that would be the farthest from Home Plate we've seen this upper Rogan unit, and of course knowing its orientation and the attitude of the bedding would be rather critical to understanding how Home Plate was basically formed in the first place, so there's basically a crater formed and with a rim of Roganlike material, or whether there's a Rogan material draped over a crater that's formed, there's also the nature of that unconformity between the Rogan unit and the underlying material. So seeing that up close would be really useful to do.[27]

Joseph's annotated image presented both what was known about Home Plate (using strokes of color overlaid on the image to identify regional units) and what was not known (using question marks). As he told it, the point of annotating these images was to direct attention to unknown features and provide the context for observations that could test his hypothesis (according to which the landscape was colored) about the distribution of the Rogan unit at Home Plate. This hypothesis relied on previous observations of this layer, believed to have been deposited while the area was hydrothermally active. Pinpointing exactly

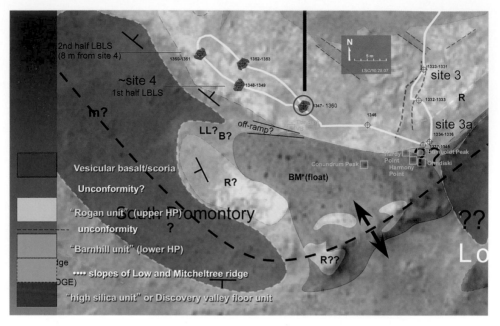

Figure 5.8. Geological map of South Promontory. Used with permission.

where else such material could be found in the region would enable the geologists to make claims about the extent, activity, and characteristics of the ancient hot spring. This would require that the rover move southward, to observe those features and fill in the blanks on the map.

Unlike Joseph's map, the areas Sarah identified as unknown were not calls to exploration but were annotated as areas that must either be avoided or be characterized more precisely with additional imaging before a safe drive could be guaranteed. The scientist's and the engineer's perspectives on the same region offered inconsistent conclusions about where the rover should drive. The same set of images therefore reveals as much about the roles and associated concerns of different team members as it does about the Martian landscape.[28] But these images also presented difficulties for coming to consensus. As Sarah's colleague Mark noted, "Of these options the north side of Home Plate is from the engineering perspective the better choice, but I understand this is not just an engineering decision."

Political implications were also deeply considered as scientists on the line debated whether they should try to move south, which meant they could "continue to explore and not retreat to places we've been before," or whether such a move signaled a "transition from bold to suicidal." But the rover's failing capacities remained the primary consideration. "We're talking about climbing ten- to fifteen-degree slopes as if we know we can do it," another Rover Planner worried aloud as his scientist colleagues pored over slope maps "looking for . . . a way there [south]." "Obviously we don't want to commit suicide," a scientist assured him. "Whatever we decide as a project, we have to decide soon . . . [we have to] get moving fast," warned a SOWG Chair. With so many interpretations and considerations flying around and pressure to "decide soon" mounting, a team member finally got exasperated with the displayed HiRISE image onto which so many plans had been recorded and asked, "Can you annotate [this] in some way to indicate what's interpreted and what's real?"

At the meeting's close, many images had been presented, with many possible plans and visions of the Martian terrain, but no consensus had emerged as to what the rover should do. The region had been well characterized by constituent groups, but this very fact produced tension over what to do next: how to move to the next step. There was no time left to delay by taking more images to better inform the decision, to develop consensus through back-channel talk, or to enable passive agreement. The mission's Project Manager and the Principal Investigator therefore convened at JPL to review all the presentations and come up with a decision. After "an agonizing evaluation," as the Project Manager called it, at the next day's SOWG meeting he announced a decision to move north.

The decision was based on rover health and safety, mentioned in the target example as one potential appeal for solving problems. If the rover physically cannot do something, the requested observation or maneuver becomes moot, resolved through external appeal to Martian conditions. All team members are trained above all to respect the need to preserve the health of the vehicle so that the mission can continue. Appeals to impending rover death can therefore force a decision one way or another, as I will describe in chapter 6. But predictions about which maneuvers will guarantee rover health are the domain of the engineers, making Sarah's map the key factor in the decision: not Joseph's or anyone else's. The team's overall anxiety in this moment was due to their inability to find common ground between scientific and engineering interpretations of the images and thus not being able to proceed with consensus derived from collective agreement.

Some of the scientists were therefore dismayed at how the decision was made. After the decision was announced, one scientist spoke up to question it, asking how it had been decided and why, and suggesting that the science team had been "railroaded" into a decision. What this scientist seemed to object to the most was not being heard in the final analysis and having to go along with a plan that not everyone had agreed to. The accusation here was leveled against the team norm of consensus. But the PI had also explicitly invoked the scientists' opinions at the End of Sol meeting, even saying that Sarah's findings were preliminary and should in no way influence their presentation of scientific rationales, which conformed to the team norm of listening (even though Sarah's map could trump all in the end). As another scientist resignedly put it in response to the decision based on rover survival, "Reality sucks sometimes." The perceived reality here was dual: both the reality of *Spirit*'s threat as Sarah had depicted with the lily pad map and the reality of the final decision. But neither perception was *seen as* reality until an interpreted image was implemented across the team as the vision for the rover's next steps.

Conclusion

Examining moments of disagreement and eventual coordination over visual interpretation demonstrates that image annotations and other manipulations must be understood in the context of social relations. Images may be annotated or processed by individuals, but they are also submitted to and discussed by the team so that they come to represent a collective interpretation. These images do not simply record a group's epistemic commitments about Mars, denote subgroups' perspectives on various aspects of the Martian surface, or serve as translational documents between distinct disciplinary groups. They also represent a social achievement within a micropolitical system that ultimately informs that epistemology. Subsequent images or interactions may be enrolled within that micropolitics as well, as the team uses images, annotations, and talk about images to quell disagreements and minimize internal divisions. Even as they effect Martian interactions, the annotated images that persist in LTP reports, End of Sol presentations, and Team Meeting slides stand as a token of moments of agreement, reminding the scientists of their belief that it is the collective and cooperative nature of their vision that will guarantee the best possible decision making and the best possible science on Mars. *Drawing* Mars *as* anything at all requires drawing the team together as well.

It is extraordinary to witness an image whose representational quality is judged on how well it represents the group that constructs it, not only the object it purports to represent. While we tend to think of representations as standing between an observer and the world, they also represent an observer's work in the world. Images on the Rover mission are drawn as a representation of a hypothesis or an interaction, to be sure, but they are also crafted in such a way as to generate or otherwise require a shared vision within the team. Annotating images is a practice through which MER scientists fashion themselves as members of a collective that demands particular social conduct and particular norms. The externalized retina[29] produced through these images is not only graphic and spatial, but also collective: the rover's images are, after all, interpreted and animated by a team. The Mars Rover mission's work flow and distributed team may make this especially visible, but I submit that the finding ought to hold for other representational work in scientific practice as well.

With this in mind, we might address the politics and social relations of map-making on Mars. Rather than taking a macro approach to analyzing the politics that enroll state actors and colonial engagements in producing maps of unexplored terrain, a subtle point is evident at the micro scale: that images reflect and project the local social orders and social relations that produce them. In this case, images of Mars are produced through the peculiar and local arrangements of producing consensus. Since they are *drawn as* a representation of that social order, rover images are *seen as* representations of consensus as well. Internal to the mission, they are invoked and produced as documents that represent consensus achieved again and again at each consecutive point, reaffirming the group's collective orientation. And while members of the Rover collective require and produce distinctive visions of the landscape, they do so only to flatten and eradicate those very distinctions in the production of their unified stance on Mars: a stance that is not only figurative but, as I will show in chapter 6, embodied as well.

Visualization, Embodiment, and Social Order

"My Body Is Always the Rover"

I am sitting in a windowless room at one of the universities affiliated with the Rover mission, next to Liz, who is staring intently at her screen. Liz is one of the Pancam operators. Unlike the calibrators, who are far from the action of the mission's daily work, Liz attends SOWG meetings as a PUL (Pancam Uplink Lead), asks questions of the scientists who suggest images, then codes those scientists' image requests for daily upload to *Spirit* and *Opportunity*. On Liz's screen is a simulated Mars rover field of view, assembled from black-and-white low-resolution Navcam images. She must use these images to indicate how and where the Pancams will point to take high-resolution color pictures. The image she is planning now requires her to command the rover to look downward and take a close-up picture of a rock between its front two wheels.

Liz looks at her screen and tilts her head to one side for a moment. Then she takes her cell phone out of her purse and places it on her desk in front of the screen. She raises her hands to the sides of her head, forearms straight, head tilted slightly downward, fingers lightly curled (fig. 6.1). Slowly, almost mechanically, she twists side-

Figure 6.1. The bodywork of image planning. Camera operator uses her hands to approximate the location of the Pancam's eyes and uses her cell phone to model the location of a rock she wants the rover to image on Mars. Drawing (from author's photo) by Craig Sylvester. Used with permission.

ways from the waist, mindful of the location of her hands relative to her phone. When I ask what she is doing, she explains: "So that's [points to her cell phone on the desk] 'close-up rock,' and then I know there's a disconnect [raises hands to each side of her face] between left and right eyes. So I have to move my head like this [tilts her head down, rotates at the waist, tilting right hand higher than left], and I have my left eye here [pauses], and then this [swivels to the opposite side, keeping head down, with left hand higher than right] is my view of the right eye."[1]

Then, in a continuous association of speech and gesture, she demonstrates for me how she associates her body with the rover's, piece by piece:

> My body, by the way, is always the rover. So right here [touches chest] is the front of the rover, my magnets are right here [raises head, touches base of her neck], and my shoulders [touches shoulders] are the front of the solar panels, and that's [leans forward, splays both arms out behind her at forty-five degrees (fig. 6.2)] the rest of it. So I have all kinds of things [antennae] sticking up over here [gestures to back], um [laughs]. But when I'm taking a picture of something in the atmosphere, then it helps me to kind of look up [looks up and sits up straighter], being the rover, and this is the front of me [touches chest] and then I put my head up [puts head up, looks back and forth] wherever, to whichever vector I'm looking at.[2]

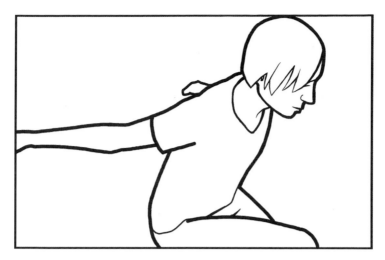

Figure 6.2. "And that's the rest of it." Embodying the rover's solar panels and bodily stance. Drawing (from author's video) by Craig Sylvester. Used with permission.

Visualization involves more than eyes and hands. As scholars of visualization technologies have emphasized, bodies are involved in visual interpretation as well.[3] Consistent with their accounts, in this chapter I will articulate the embodied practices of image interpretation involved in working with rover images. I will further elaborate the role of these embodied sensitivities in team members' understandings and projections of robotic experience, and their role in producing team solidarity.

In doing so, I make an analytical move from *drawing as* and *seeing as* to *seeing like*: specifically, seeing like a Rover. The three are interrelated. Seeing like a Rover is the result of *drawing as* techniques that produce and transmit a particular sensitivity to rover vision and mobility. Placing rocks and other driving hazards in the perceptual foreground, the aspect presented in these images shows Mars as it is (or could be) experienced by the rovers.[4] Many of these *drawing as* practices were discussed in chapter 4 as the mission's maps, or in chapter 5 as part of producing a consensus view of Mars. Here, however, I want to emphasize how such visualizations and their associated talk and gesture reproduce social order by developing an embodied attunement to the rover's experiences on Mars, which team members vicariously experience on Earth. I argue that these visualization practices, their associated gestures, and the embodied narratives are also organizational practices that produce and maintain the mission's local, consensus-based order.

Figure 6.3. Hazcam image from the lab, prelaunch. Image credit NASA/JPL/Caltech.

The Aspect of Trafficability

A new team member's introduction to learning to see like a Rover is often exposure to Hazard Avoidance Camera (Hazcam) images. The four Hazcams,
mounted under the rover deck and facing downward between the rover's front
and back wheels, have a fish-eye lens that enables the robots to capture a broad
view of the horizon, up to 120 degrees (fig. 6.3). This particular distortion enables the rovers and their human team members to see a broader range of the
nearby environment, making it easier to identify hazards or zones for interaction. Correcting a fish-eye photograph to a rectangular frame is easily done in
most image-processing software suites. But rather than drawing the image in
ways conventional for a human observer, scientists and engineers frequently
work with these images in raw form. They speak instead of adapting their eyes to
this particular way of viewing the Martian surface. In the course of my research,
many scientists gave me different explanations of how one should acquire this
aspect. One referred to a preflight photograph taken with the same lens (fig.
6.3) that helped him learn "how to see" with the Hazcams (fig. 6.4): "For me, I
need pictures like this [points to a Hazcam photo of people in a lab on Earth] to

Figure 6.4. Hazcam image of Mars, with curvature from fish-eye lens. Image credit NASA/JPL/Caltech.

make the correction . . . this [points to a Hazcam image from Mars] sort of looks normal, but it's being warped and distorted."[5]

When Hazcam images are displayed in a SOWG or End of Sol meeting, scientists often remind their colleagues that the optics are distorted: as Roger often jokingly puts it, "objects in the mirror are closer than they appear." But this reminder is not so much a caveat as an invocation of shared knowledge and tacit skill. Rover scientists also share a (perhaps apocryphal) story about a reporter for a major newspaper who, on seeing the Hazcam images posted online at the JPL website shortly after *Spirit*'s arrival, publicly commented that Mars had a sharper curvature than Earth. Such accounts point to how developing and invoking others' visual expertise with the Hazcams is a way of identifying a fellow team member who maintains the same intuition for the rover's-eye view of Mars.

Familiarity with software on Earth and on Mars also plays a role in acquiring this aspect. The robots are equipped with onboard artificial intelligence capabilities to analyze Hazcam images and evaluate whether a rock in its path is too large to drive over. If it is, the rover can modify its course somewhat to avoid the hazard, overriding the instructions sent by its human operators and driving around the object. As they drive, then, the vehicles must periodically take pictures and analyze them before moving ahead. As Mark, a Rover Planner, put it:

> For one thing, the rover's view of the world when driving is very much like your view of the world if you imagine yourself trying to make your way through a dark, cluttered room with nothing but a flashbulb. So you can kind of take a picture in the world, and you can get a sense of where there's a safe path, and you walk a little way along that safe path and you pop the flashbulb again. . . . That's one of the ways the rover sees the world when it's driving. Other times it just does this [he throws his hands in the air]: "All right, I'm going to just go where you [Rover Planners] tell me."[6]

Just as the rovers' software actively looks out for hazardous elements of the terrain in order to safely execute driving instructions, Rover Planners adopt a parallel sensitivity to the Martian terrain, even as they are responsible for coding those very instructions. These specialist engineers are particularly adept at identifying, in the images that return to them, rocks, slippery soil, sand traps, and other potential obstacles that would be likely to trip up a five-foot-tall, six-wheeled rover out in the wilds of Mars.

Team members use software on Earth to visually construe Mars so as to produce this aspect. Digital elevation maps (DEM) are imported directly into the rover planning tools, so that operators can not only see what it looks like around their rover but also pay attention to the undulations of the terrain that might affect an upcoming maneuver. The Rover Planners most frequently draw and see Mars this way, but team members across the mission use these and similar tools daily to model how and where the rover can drive or place an instrument. Using DEM data, their software colors in patches on top of Navcam images to help team members place targets or plan an activity (cf. fig. 4.13). This not only signifies collective agreement and draws Mars as prepared for interaction, but also reinforces a view of Mars at the rover's scale.

Another essential part of developing a sensitivity to rover mobility is the ability to see two-dimensional Pancam, Navcam, or Hazcam images in three dimensions. To do this, many Rover Planners transform these images into

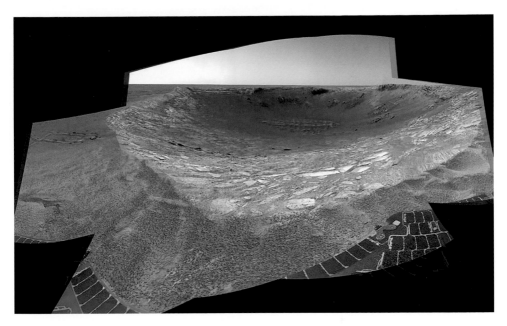

Figure 6.5. Three-dimensional view of a crater to "engage your own kinesthetic sense." Image credit NASA/JPL/Caltech.

three-dimensional projections called anaglyphs. They load two pictures of the same scene, one taken by the right camera and the other taken by the left, and combine them in an image processor to craft a stereo view in which one image is colored red, the other is colored blue, and the two are offset from each other by a certain degree consistent with human stereo vision (fig. 6.5). Anaglyphs do not look like much when viewed on a screen, but once the scientist or engineer dons red/blue 3-D glasses, the scene acquires depth.

Anaglyphs are employed across the mission for various reasons. Scientists use them to get a sense of the texture or the morphology of a rock or surface feature under examination. But engineers parse anaglyph images differently. Mark explained the value of the three-dimensional view as one that engaged his kinesthetic sense for the terrain, making elements "pop out" to get a "better sense of the size and slope":

In 2-D you can't really get a sense of, Is this a big ridge? . . . [T]here's something to be said for engaging your own kinesthetic sense. . . . If you take a look at this in 3-D, you can see how it now kinda pops out at you, how this terrain is kind of undulating. . . . [W]here I could see kinda that there was a ridge here [in 2-D], this is now [in 3-D] giving me a much better sense of the size of that ridge and the slope of that ridge, and you can get a sense of there's terrain blocked behind the ridge.[7]

This language recalls Ben's description of image-processing techniques that make particular features "pop out" in false color. Mark's is a *drawing as* practice too, but here the aspect that needs to be produced is attention to robotic obstacles. The Rover Planner must see Mars as strewn with potential drive hazards to be avoided. As he verbally parsed the image with me, Mark pointed to rocks and dunes strewn across the field, "evaluating them as obstacles": "These two here are obstacles, this one here is definitely an obstacle, this stuff here is probably okay although we should stay away from them with a five-wheeled rover." Parsing an image in this way, this Rover Planner not only demonstrates his professional vision in his attunement to driving conditions,[8] he also talks through how the rover's artificial intelligence algorithms evaluate the terrain and describes his own understanding of how the rover would need to interact with the field and how to keep the vehicle safe. Describing the terrain in this way and verbalizing what he can see, Mark demonstrates his expertise at seeing like a Rover.[9]

Importantly, then, seeing like a Rover requires more than drawing Mars as a map, as tangible and interactable. It requires drawing Mars as tangible and interactable *for the rover*. It requires knowing how the rover photographs, moves, navigates, and interacts with the terrain in order to craft visualizations that show where and how the rover can drive: what the team calls rover trafficability. In the foreground, along with rocks, hills, dunes, and other drive hazards, is the team's knowledge of the possibilities and limitations of the robotic body. Before they can competently make decisions about how and where to drive or program their instruments to conduct an observation, then, team members must learn how to see Mars from this frame of reference. This includes acquiring expertise about the rovers' visual apparatus, as in the case of the Hazcams. But seeing like a Rover also includes developing a parallel intuition for the rover's body and its mobility: both its visual and its bodily apparatus.

Embodiment: Gesture and Narrative

Seeing like a Rover, then, enrolls visual practices to cultivate an embodied sensitivity to the robot's interactions with Mars. Note that not only Mark's vision but also his "kinesthetic sense" is at play. Here, "engaging your own kinesthetic sense" is a question not of projecting human kinesthesia onto the robot, but of adopting the robot's sensitivities and mobilities. Certainly the robots are subject to a degree of anthropomorphism, since parts of the rover are verbally related to human body parts and actions. For example, the Pancams are regularly referred to as the rovers' eyes, the hazard cameras aimed at the wheels show "what's under our feet," while the Instrument Deployment Device (IDD) is "the arm." In team parlance, the rovers talk to Earth via communication antennae, go to sleep at night, wake up and take naps at certain times, stare or look at targets on the surface regularly throughout the day. These active verbs describe technical activities but also reinforce an experiential dimension of these activities consistent with human experience. But in many visual, gestural, and narrative moments entailing the rovers, the projection does not run from human to robot, with the robots acquiring human characteristics. Instead, individuals on the mission must learn, imitate, and demonstrate what it is like to be a rover on Mars. Thus the team is subject to technomorphism as members take on the robot's body and experiences in their accounts of their work.[10] This is performed through particular forms of talk and gesture that write the rover onto the human body.

A key aspect of adopting the rover's "kinesthetic sense" is a developing a sensibility to what the rover might see, think, or feel related to specific activities that must be planned. Pancam operators are highly attuned to the sun's position relative to Mars throughout the day, attributing their heightened sense of Martian light and shade to knowing how to see with the rover's eyes but also to knowing where the rover is and if its shadow will be visible in the photographic frame. A RAT (Rock Abrasion Tool) operator talked about his instrument as the rover's "sense of touch," describing the output graphs of drill intensity as descriptions of how the rover "feels out the rock."[11] Mark confessed to me that when planning a drive, "I have frequently tried to put myself in the rover's head and say, What do I know about the world?" He then elaborated by describing the differences between himself and the rover: "The rover has senses that we don't have . . . the rover sees stuff that we don't see, it sees into wavelengths that we don't see, it never really sees the world in color but it can see parts of the spectrum that we can't."[12]

Jordan, another rover driver, also related his sensitivity to the rover's experience to his own physical bodily sensations, in terms of feeling and intuition based on prior action. For Jordan, working with the rovers requires "having a feeling" about the robot's present and upcoming activities, much the way he would cultivate similar feelings about his own body based on his actions:

You just have more of an intuition as to, I think, I don't know if this is a good example or not, but you know as you get older you understand how your body works more and so you know the effects of, if you haven't eaten breakfast or something, you know by lunchtime you can feel . . . you know why you feel differently right before lunch as opposed to yesterday when you had breakfast. And so operating the vehicles after a while you get an idea of well, okay, the rover did this yesterday so I have a feeling, I know what it's going to be like tomorrow. Or I know it did a really long drive yesterday, so I have a feeling.[13]

Enhancing this intuitive and embodied connection are a set of visual and material practices that, taken together, become a kind of physical calculus for working through rover motions and activities on Mars from Earth. Mark developed a set of paperweights that mimicked the degrees of force the rover could use on Mars, to better bring the rover's experience into his colleagues' bodies on Earth. Ben keeps a piece of paper cut out in the shape of a Pancam frame that he lays over his screen to get a sense of the Pancam's field of view: what a proposed observation will include. Pancam operator Jude recalls how she and her colleagues "used to put Post-It notes on our foreheads so we could know how the [Pancam] frames would turn out." The Pancam software requires its users to place digital yellow squares on top of Navcam images to indicate where the rover will take its next Pancam image (cf. fig. 4.14). These operators imagined the Post-Its projecting from their foreheads, through the corresponding yellow squares on their screens in the command software and from there projecting outward into the Martian terrain.[14]

Gestures serve as a physical calculus too. Liz's elaborate bodywork described at the outset of the chapter is an example of a codified suite of gestures that are common practice across the team. Team members regularly manipulate their shoulders, elbows, and wrists to mimic the robots' range of motion, and when estimating their position they splay their arms out to both sides to imitate solar panels and tilt their bodies to approximate the rover's pitch and yaw. One of the most common gestures I observed on the mission is using one's own arm to demonstrate how the rover deploys its IDD, informally called its arm: this involves lifting the right upper arm to shoulder height, dropping the forearm to ninety degrees with the fist pointed at the ground, and articulating the arm in a limited fashion first side-to-side from the shoulder, then swinging forward from the elbow. As Mark explained, "When we're training new rover drivers, we can really tell that they get it when you start talking about moves with the IDD and they start moving their own arm to kind of show you what they mean, and they say, you know we're gonna swing this to the left and then move their elbow [moves his elbow to the left, wrist hanging down]."[15]

He confessed that he and his colleagues "used to talk about how the rover was going to go by scooting around in our chairs." I too witnessed a SOWG meeting where a scientist proposed a new maneuver and another scientist in the room used his wheelie chair to work through the move as it was being described.[16]

It is tempting to analyze these gestures as communicative acts: ways of trans-lating the skill of embodied digital seeing from one team member to another.[17] I certainly observed situations where this was the case, where a wheelie chair maneuver or a skilled twist of the elbow was a central articulation in the work of communicating and coordinating action at a distance between team members on Earth. However, most of the times I witnessed these gestures, the interlocu-tors were not in the same room. Most frequently scientists, engineers, and tech-nicians alike gestured in what were clearly formal, codified, standardized ways of enacting the rover, but they did so while alone, speaking to mutually invisible meeting participants on a teleconference line.

Bodily sensitivity to the rovers' capabilities changes as the rovers degrade or perform new feats. Some of these feats are improvisational, relying on this very bodywork to inspire new maneuvers that the rovers were not originally built to do. I have witnessed members working through robotic activities with their feet, arms, and eyes to suggest digging trenches with the rovers' wheels, or using the Micro-scopic Imager to take a picture of a problematic component on the Pancam mast or the underside of the rover's body. As parts of the rovers break down, the team's bodily sensitivities adjust accordingly. When *Spirit*'s right front wheel stopped working, the Rover Planners started to drive the robot backward, dragging the stuck wheel behind. In this make-do arrangement, the rover serendipitously turfed up the white soil at Tyrone. Within days of this discovery team members stopped referring to "our crippled rover" and started calling the bum wheel "our furrowing tool."[18]

Another kind of narrative common across the team similarly draws an inten-sive connection between members' bodies and those of the rovers. Countless team members I interviewed explained that their eyes have "become Pancam, or Navcam."[19] Another put it simply: When working with the rover, "I am a rover. I am a Pancam."[20] Other stories assumed a somatic, even causal, association be-tween the robot's experiences on Mars and their own bodies. As a midcareer female scientist recounted, "I was working in the garden one day, and all of a sud-den I don't know what's going on with my right wrist, I cannot move it—out of nowhere! I get here [to the SOWG meeting], and *Spirit* has, its right front wheel is stuck! Things like that, you know? . . . I am totally connected to that gal!"[21]

Matter-of-fact statements like these are not limited by age or gender but occur across the mission. Here is another example, from a young male engineer: "Interestingly, I screwed up my shoulder . . . and needed surgery on it right about the time that *Opportunity*'s IDD [arm] started having problems [with a stiff shoulder joint], and I broke my toe right before *Spirit*'s wheel [broke], so I'm just saying, maybe it's kind of sympathetic, I don't know [laughs]. I mean, I don't think there's any magic involved or anything, but maybe it's some kind of subconscious thing. I don't know."[22]

When the rovers are "healthy" or "sick," human team members on Earth may exude energy or tense up. Jude explained to me that when something is not right with the rover, "We feel it in our bodies." During the dust storm in the summer of 2007, team members were very much on edge, perceptibly anxious about whether their rovers would survive. Liz articulated a comparison drawn by several team members: "It's like if your grandparent is sick and in the hospital and there's nothing you can do about it. You just have to trust that the doctors are doing all they can."[23]

As an ethnographer on the mission, I too learned these members' methods. I could tell whether it was a good day or a bad day on Mars based on the PI's footfalls in the hallway outside the meeting room, or in the slouch or spring in his colleagues' steps in the lab. I also started to feel the curious different bodily experience of working with *Spirit* as opposed to with *Opportunity*. It is difficult to express verbally, but I can feel it in my body: a kind of posture or stiffness associated with each robot's different faults and features, like *Spirit*'s stuck wheel and *Opportunity*'s stiff shoulder joint. I acquired this skill slowly over my first six months of fieldwork, at the same time that I was learning to see and understand team members' representations of Mars. When I finally came face to face with a rover for the first time a year later (the one kept on Earth to try out new commands before they are sent to Mars), I felt as startled and awkward as if my own reflection in a mirror had just extended its hand to me. Taking on a different bodily sensibility is a disembodying and disorienting experience, but it is also one that comes with the process of acquiring membership and accumulating members' visual skills.[24]

Visualization and Embodiment

Robotic bodies, gestures, and ailments seem to bring us a long way from visualization. But recent scholarship on technologies of visualization has argued that we should consider bodies an essential part of visual practice. Sociologist of science Natasha Myers calls these gestures "bodywork" and describes them as practices that accompany expertise in molecular biology modeling: "As [the scientist] tells the story, she contorts her entire body into the shape of the misfolded protein. With one arm bent over above her head, another wrapping around the front of her body, her neck crooked to the side, and her body twisting, she expresses the strain felt by the misshapen protein model."[25]

Although the biologists Myers studied used their bodies to develop an intuition for the simulation's accuracy, ethnographer and semiotician Morana Alač examined the use of gesture to make sense of brain scans displayed on screens. In Alač's account, following research scientists at their computers, such gestures are semiotic acts, much like speech: they must be coded and understood alongside talk, text, and images as essential to making sense of digital fMRI scans. In both cases, bodily activities are part of the process of interacting with visualizations to make sense of the depicted protein or brain.[26]

Embodied gesture and narrative on the Rover team are also important for making sense of Mars, although not necessarily by enacting Martian features. Another fruitful line of inquiry here is that offered by medical anthropologist Rachel Prentice in her study of minimally invasive surgery. Showing how surgeons "see with their bodies," learning their way around tissues and organs without direct sight, Prentice explains how tools such as scopes extend this bodily seeing. The surgeons she observed and interviewed described how they imagine their eyes located at the point of the scope's camera, deep in the patient's body, or how the tools becomes extensions of their eyes or fingers. She cites a surgeon who, while operating on an arthritic shoulder and looking at the screen where the video feed is projected, says, "Actually I would say I am sitting on that piece of anatomy, or rather that you are floating around, swimming around in the [joint]."[27]

Central to Prentice's work is that of phenomenologist Maurice Merleau-Ponty. Prentice appeals to Merleau-Ponty's notion of "proxies" that augment and extend our bodies out into the world[28] to support the assertion that visualization—even instrumentally assisted visualization—is always situated and embodied. Instruments do not somehow render the body neutral or objectively compose and transmit the world to the viewer. Rather, like a blind man's stick, they move our point of perception from the point of interaction between our hands and our instruments to the point of interaction between our instruments and the world.

There is a tempting parallel here to Rover team members' accounts, many of which also use "proxy" to describe their rovers on Mars. Their use of the term explains how the robots go where humans cannot go, see what humans cannot see, and allow humans to "be there with them" on Mars. Some analysts have used this emic language analytically to describe the rover as an extension of the team on Mars that allows team members to forge a novel identity as field geologists who work with robots.[29] After several years of immersion with the team, however, I am reluctant to convert this actor's category into an analytical one. The language of the proxy is most frequently used in discussions with those who are not team members, such as the press or interviewers, but it is not visible at the point of practice. It may indeed be that the proxy as projection of self onto Mars is a common form of talk recognized across the team, one that articulates the intimacy of their relationship with the vehicles and the status of their own agency in producing this instrumentally assisted vision. But what I witnessed team members doing while working with rover images was quite different.

To understand the images of Mars that the rovers return, team members do not project themselves outward, into the body of the rover as human proxy. Rather, they themselves adopt the rover's bodily apparatus with its unique bodily sensitivities in order to understand and interact with Mars. As Liz explained,

> In order to be fully prepared for my job . . . I need to literally *be* that vehicle. That's what all the visualization software I use is about. . . . For me, it's all about intuitively being able to make decisions, because you're gonna be getting questions on the fly and you're gonna have to answer them on the fly. You're not you, you're the rover. . . . You're thinking for the vehicle. . . . I think of myself as the rover so I can call the shots. I need to know where I am as the rover. It's a huge, huge part of my job.[30]

Liz here associates her visualization software and image-processing practices with her daily ability to "literally *be* that vehicle." Roger also explained to me in our interview that while the rovers' own "Athena payload is [the] embodiment of a geologist on the Earth," in the context of daily operations one must always "think like you're in the body of the rover."[31] The vehicle may be called a robot geologist, but in its approximation of a human geologist's skills and tools, it reconfigures the practices of looking, moving, and conducting experiments such that team members must radically adopt their bodies to its frame of experience. So it is not only that the rovers are the human team members' proxies on Mars: rather, the visual and embodied practices described above show how human team members must step into the rovers' bodies in order to experience Mars.[32] The proxy, then, does not run one way: embodiment is a two-way street.

Embodied seeing, as Merleau-Ponty articulates it, also produces a *seeing as* experience. As he puts it, "It is necessary to put the surroundings in abeyance the better to see the object, and to lose in background what one gains in focal figure . . . because objects form a system in which one cannot show itself without concealing others."[33] The body's senses and mobility within an environment are an essential part of this perceptive practice: that is, we compose the world as we move and see within it. This resonates with philosopher of science Hans Radder's connection between theory-laden observation and embodied inter-action. For Radder, the perceiver's movements and actions also compose per-ception alongside concepts, language, and theory, even "over and above . . . the instrumentally embodied extension of human sense organs."[34] Radder calls this "material realization." The perception of material objects involves all the human activities that make such observations possible, such as working with instru-ments, organisms, or institutions to conduct scientific observation. So bodily interactions can also structure observational processes such that they produce different *seeing as* experiences. As he puts it, "Any observational process is always materially realized and conceptually interpreted right from the start."

The embodied practices and narratives I describe in this chapter are there-fore essential to visualization, perception, and sense-making with rover materi-als. The team's experience of the rover's body is implicated in observations "right from the start."[35] Narratives, gestures, and visualization practices that *draw* Mars *as* the rover might encounter it produce a shared, embodied sense of the rover's bodily apparatus. This enables team members to see like a Rover, to make sense of the images the robots return, to plan for future observations and interactions, and to compose their visualizations on Earth. Drawing a connection between human bodies and distant robot bodies is essential to making knowledge of Mars.

Importantly, these codified gestures and narratives do not allow humans to actually become or become like their rovers. They are members' knowledge: rituals, practical accounts, and performances that are learned as part of join-ing and being part of the team. One cannot move one's arm any which way to behave like the rover: there is a right way that members know, by which they recognize their own. Knowing how to move like the rover is also a question of belonging. Recall the interviewee who could "really tell when [new members] get it" because of the way they moved their arms. "Getting it" is not just "getting" how the rover works, but also "getting it" in terms of participating in the team's recognized gestural practices. This reveals a previously unexplored aspect of em-bodied talk, gesture, and visualization: as practices that construct and maintain social order.

From Embodiment to Social Order

A few weeks after Liz performed her rover gestures at her desk, she described the rover to me as "the glue that bonds the team together," especially as it moved to a distant location and had to be invoked (or appresented, in Knorr-Cetina and Bruegger's terms)[36] daily:

> The hardware [a rover] is like the glue that bonds the team together while it's being built on Earth. During that time, we can directly relate over something physical. Once that spacecraft is off the ground, that connection moves into the software realm, and also into our minds. So I'd argue that the dynamics of the team took on greater meaning once *Spirit* and *Opportunity* left the planet. Once those rovers leave Earth, the team is all we've got.[37]

In Liz's repeated invocation of "the team" alongside the robot's hardware and software systems, she discursively aligns the distant bodies of the rovers not with herself as an individual, but with team dynamics. Mark also crafted the same intensive connection between the team's collaborative practices and the rovers' activities: "We are all, like, working together. We are the corpus, the body of this rover. We are making that thing do what it does on Mars."[38]

In these accounts the rover's body emerges as a representation of the team. Although constructed as a distant teammate,[39] it reflects and supports a particular kind of community alignment: in this case the politics of consensus. An analytical and an analogical touchstone are instructive here to support this position.

The former derives from the foundational sociologist Émile Durkheim's description of solidarity and social order as produced through ritual practices. Drawing on anthropological literature of the day, Durkheim characterized "elementary" religions as concerned with the management of totem animals, plants, or other protective forces.[40] According to Durkheim, totems serve structural functions in their societies, since their characteristics and associated rituals assert the local culture's categories and structures, such as social hierarchies or divisions between the sacred and the profane. Care of the totem requires adherence to elaborate rituals that perform the social order of the group and gathering in "effervescent" assemblies that include dancing or other gestures in which members of the group may imitate the object that brings them together. As Durkheim puts it: "The totem is their rallying sign; . . . it is no less natural that they should seek to resemble it in their gestures, their cries, their attitude. . . . By this means, they mutually show one another that they are all members of the same moral community and they become conscious of the kinship uniting them. The rite does not limit itself to expressing this kinship; it makes or remakes it."[41]

Recall the careful ritual structure of the SOWG meeting, the repeated emphasis on solidarity and unity through consensus building, and the ritual effervescent assertions of "happiness." I suggest we approach the series of gestures and activities that unite the rovers' bodies with the bodies of their team members as ones that build and maintain a totemic relation between team members and their robots, thereby cementing social relations between those team members. The intensity of the emotional connection to the robot is heightened, thus continually reinforcing the emotional energy of the team.[42] Bound up in producing legible images of Mars, then, are embodied visual practices that produce a team in its particular collectivist social form.[43]

Durkheim's perspective is central to sociologist Randall Collins's notion of interaction ritual chains. According to Collins, interaction rituals and continuous conformity to them produce the "emotional energy" that binds a group together as a unit. Thus Durkheim's cohesive societies appear to have "mechanical solidarity" owing to their high social density (copresence) along with their low social diversity (localism). Intensity of emotion produced through shared ritual reinforces centralized nodes of participation and low degrees of social stratification. The rituals of daily SOWG meetings, consensus formation, "happiness," and embodied subjectivity experienced through team members' embodiment of the rover on Mars may also produce this "emotional energy" that binds the group together as a single team, copresent and located on Mars, within a flattened hierarchical social form.

While Durkheim and Collins give us tools to think analytically about these practices, a good analogy is that of the body politic, especially as depicted by Thomas Hobbes. In his frontispiece to the *Leviathan* of 1660, Hobbes drew the king's body as a "body politick": that is, composed of his many subjects who together grant him authority for action on their behalf, making him the royal "we" (fig. 6.6). The rover's body is a "body politick" too. Team members unite to grant the robot the authority to act on their behalf and are complicit in the robot's activities. The rover's authority for action rests on the team's orderly sociopolitical structure: in this case, the groups' collectivist orientation and the continued production of unilateral consensus. Rover activities and rover agency are predicated on this internal organization, expressed through the robot's body.

This is especially evident in talk that accompanies images or image work, especially those visualizations that draw Mars as trafficable for the rover. Other observers of the Mars Rover mission have noted team members' predominant

Figure 6.6. Frontispiece to Thomas Hobbes, *Leviathan*, showing the king's body composed of his subjects' bodies. Princeton University Library.

use of "we."[44] In my own ethnographic experience too, team members substitute the first-person plural pronoun for the rover when planning its activities on Mars. For example, at the outset of SOWG meetings, the LTP Lead will frequently put an image on-screen (fig. 6.7) and talk about it thus: "We expect to turn around and take images of [the target]. . . . We're about four meters from the outcrop that we wanted to image, and so the idea was to bump forward maybe two or three meters so we can get better images and MiniTES observations."[45]

The use of "we" refers to the team and the rover, subtly aligning the team, with its interests or actions, alongside the actions of the robot. Not unlike the royal we, which aligns monarch and subjects as a composite actor, the pronoun binds the team together into a relationship complicit with its rover's activities. For another example, return again to Sarah's End of Sol presentation about using South Promontory as a winter haven, discussed in chapter 5. Her language is peppered throughout with a "we" that conflates the Rover Planners, their local drawing as practices, and even the rover itself.[46]

Linguistic anthropologist Elinor Ochs has noted that other scientists have a tendency to place the subject of the sentence in the position of the object of study. The physicists she studies make statements about particles such as, for example, "When I come down, I'm in the domain state."[47] But using "we" along-side rover images goes a step further, situating a collective behind the rover's eyes and producing an intersubjective position that effaces the individual. Another useful point of reference is sociologist Emanuel Schegloff's description of forms of talk that formulate place. As I have described in previous chapters, such utterances establish a shared geographical location and point of departure and identify membership within a conversation.[48] In each of the examples above, however, the use of "we" is punctuated with interpreted visions of the Martian terrain. Thus these forms of talk and associated visual practices — *drawing* Mars *as* trafficable for the rover, using the pronoun "we" — orient the team within the rover's frame of reference to establish a shared subjective position, located in the body of the rover on Mars. At the same time, they identify all participants on the line as engaged in the same collective process.

"We" is invoked even when there is some disagreement over where the rover should go and what it should do. Recall how Sarah accompanied her appraisal of Winter Haven 3 with slope maps, lily pad maps, and other visualizations: in doing so, she not only *drew* Mars *as* trafficable for the rover, but also inspired her teammates to adopt her perspective as the rover's own, and to therefore *see* South Promontory *as* impassable. Similarly, shortly before the Winter Haven 3 discussion, when advocating that the rover move to examine more silica-rich materials to the southeast of Home Plate, Nick, a scientist, attempted his own visual analysis of the terrain. Presenting an image (fig. 6.8) that he had marked up to suggest a "minimal ridge" (a trafficable drive route), Nick addressed Kwame, the Rover Planner on duty that day: "Based on your [the Rover Planner's] presentation the other day, you showed one on-ramp [to Home Plate] that does look unapproachable or difficult, but what I was trying to show with these images . . . was an alternative that I wondered if you guys have looked at as well."[49]

Kwame countered that the Rover Planners had already "looked at all the southern approaches [to Home Plate], and we don't think they're viable. We don't wanna get stuck somewhere we cannot recover."[50]

Note that while this begins as a potential question of pitting a scientist's skills at drawing Mars as trafficable against an engineer's skills, Kwame then invokes the rover "we": "We don't wanna get stuck somewhere we cannot recover." It's elusive here whether his use of "we" refers to *Spirit*, the Rover Planners, the team in general, or some combination thereof. This is an effective way to resolve dispute by formulating visual interpretation as intersubjective, experienced from the rover's-eye view. To resolve the conflict at hand, a second scientist jumped

Figure 6.7. A slide from an LTP report at the outset of an *Opportunity* SOWG meeting before the Victoria Crater approach, giving a rover's-eye view of the terrain parsed for collective presence and orientation within the rover's body. SOWG meeting, sol 944, September 19, 2006. Used with permission. Compare with figures 1.2 and 4.4 from the same planning period.

in to suggest a third option, a middle ground that both was "reachable" for the rover and "looks smooth." This shifted the conversation away from a conflict of professional visions while continuing to parse the terrain as trafficable for the rover, seeking out an area that all would agree was safe for the robot to explore.

Rover Death

The two rovers were designed to last ninety days each. *Spirit* and *Opportunity* both "outlasted their warranty," as team members describe it, by over two thousand Martian sols.[51] Far from producing complacency, during my fieldwork the potential death of the rovers was evoked as a constant threat. New team members were acculturated into the belief that the rovers were finite resources and that their short lives could be over at any time. This was colloquially referred to as "the sniper": as in the expression "the sniper could strike at any time." This produced a particular urgency for whatever observations were immediately planned. But it was also invoked as a way of managing team members on the mission, bringing them into alignment with orderly expectations and totemic relations. Just as Durkheim describes the loss of the group totem as "a disaster . . .

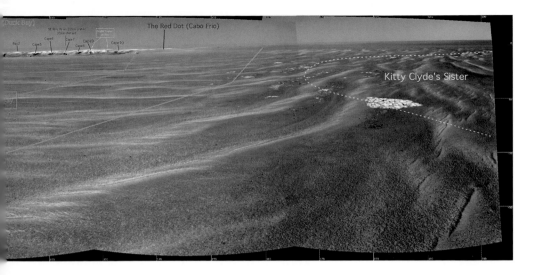

Figure 6.8. Nick's use of false color and annotation to "see like a Rover." Note how this false color combination makes drive-hazardous rocks "pop out," and note the use of the line and text to indicate a potentially trafficable position. *Spirit*, "Home Plate Field Guide," End of Sol meeting, February 28, 2007. Used with permission.

the greatest misfortune which can happen to the group,"[52] the preservation of the totem at all costs can require individuals to sacrifice their individual interests to support the collective goals of the team, making the rhetorical appeal to rover death a supremely powerful resource in group management.

Appeals to rover death similarly draw together visualization, embodiment, and the collective social order. Recalling the examples above, such as target selection or Sarah's or Kwame's visual interpretation of rover imagery, appeals to what the rover cannot do without tremendous risk can override any scientific consideration. As the individuals who propose these expert visualizations appeal to the rover's own body and point of view, they depersonalize their suggestions and defuse any antagonism as they verbally and visually remind their teammates of the collective imperative to keep their robots alive.

This was especially visible in June 2007, when two faulty commands were uploaded to *Spirit*. The commands triggered a fault in the onboard computer, causing the robot to reset without trouble, but the mission on Earth ground to an immediate halt. The Project Manager sent all engineering team members at JPL home for an enforced four-day weekend, and SOWG meetings were canceled. On the Monday of their return, the first item on the agenda was an "All Hands Meeting," the first since the mission began in 2004. The PI described this special meeting as a chance "to return to first principles." The entire engineering team filed into the SOWG room, and the teleconference lines beeped as science team members called in from across the country. Clearly, the faulty commands signaled a significant breach in social order on the team at the same time as they signaled a potential fault on Mars.

At the outset of the meeting, the Principal Investigator and the Project Manager framed the situation as one of group responsibility, reinforcing the collectivist perspective. The Project Manager told a story about the successful Japanese auto industry, in which "anyone on the assembly line can stop the process." The PI concurred, saying, "If you see something that looks funny you are empowered just like everyone else to pull the cord, to ask questions. . . . [I]t applies to everybody as part of the process." He then advocated more procedural rigor in following the rules and the roles outlined in chapter 1: this, he believed, would return team members to the level of operations where "we're like a fighter squadron at the top of our game."[53]

After articulating these principles of the team's social order, the two opened the floor to comments from the team on what needed improvement or had gotten "sloppy" over time. Jesse, who had helped build the rovers' camera systems, spoke up to articulate a problem he witnessed with respect to the responsible use of imagery:

[At the beginning of the mission] there was sort of this culture of curiosity combined with paranoia, and everyone was on their game. . . . As people have been cycled in and out of [the team] . . . we have new people, and I kind of get the feeling that they don't have the fear [we had]. . . . It's more of a video game for a lot of people, it's kind of cool . . . it's sort of abstracted a little bit. . . . They may not be as connected to the fact that the rover is only one day away from we're never going to hear from it again. . . . [A]nything we could potentially do could end the whole game.[54]

According to Jesse, the sense of virtual, embodied presence that rover imagery inspires among a younger generation of engineers had produced a kind of affective disconnect similar to playing a video game but inappropriate for the mission. Responsible visual practice on the mission, he believed, must emphasize not only the consequences of inconsiderate play (e.g., the loss of the rover), but also the connectedness this visual mode should inspire between individuals, the rovers, and the team. The right way to approach these images, Jesse suggested, is as embodied and consequential for the "we" of the team, not as distanced and abstract. The terrifying reminder that the rover is always "only one day away from we're never going to hear from it again" is consistently imposed through visual orientation and through the embodied, collective connection that this visual orientation provides.

In this moment of breach surrounding the possible death of the rover, the visual, the embodied, the organizational, and the totemic were all simultaneously invoked to produce a corrected team orientation. Social order was very much in the foreground, evoked through the mutual entanglement of embodied connection, organizational structure, and visual modes. The PI's and Project Manager's comments articulated clearly the local norms of interaction that produce the rover "we." And Jesse argued that visual interpretation was an essential part of that process, of inheriting and enacting the team's local order. The proper way to see rover images, Jesse claimed, is to see like a Rover, with all the affective, fragile, collective, and totemic relations that implies.

Conclusion

Seeing like a Rover begins by enlisting members' *drawing as* and *seeing as* practices to develop a shared sensitivity to the "rover's-eye view." But seeing like a Rover also entails embodied gesture and narrative, forms of talk and visual practices to make sense of the rovers' experience. To be sure, this builds empathy and intimacy between team members and their distant robots. It also brings team members together in the body of the rover—an object they are all committed to—producing the team's collective complicity and solidarity through the rover's activities. The imaging, gestures, and forms of talk associated with visualization on the Rover team draw a totemic connection between team members' individual bodies and the rover's body, thereby cementing the team's social order on Earth. The practices of purposeful image construal—*drawing as, seeing as,* and ultimately *seeing like*—order both objects and subjects, on Earth and on Mars.[55]

Activities based on collective and situated knowledge, as in the rover case, can and do produce visible principles and traces of social order. As knowledge-making communities adopt or produce a singular viewpoint, this view is clearly situated in time, body, and space. After all, seeing like a Rover is certainly a question of seeing from somewhere, not producing a view from nowhere.[56] It is from this situated perspective, the rover's-eye view produced and maintained through *drawing as* practices and practiced through embodied talk and gesture, that members are socialized into a collective seeing that requires them to work together with each other and with their robots. As rover teammates acquire the members' techniques and accounts of seeing like a Rover, learning to place their own bodies into the bodies of their machines with their associated stiffness, visual modes, and frailty, they both perform their group's membership and produce the radical collectivity that constitutes the social order of the team.

AGENCY AT THE INTERFACE: THE "WE" AND THE "SHE"

Dusk is falling over the field at the Caltech Athenaeum, where team members have gathered to remember *Spirit*. The conversation is lively for a robot funeral. Mark and his scientist colleague Adam are engaged in an animated back and forth about just what, or who, *Spirit* is. From my time with the team, I'm certainly familiar with how the rovers have different personalities, and even social class. As Mark once explained to me,

> *Opportunity*'s sort of the glamour girl. She went to Mars to find water, and she sort of fell into a hole and opened her eyes and there's evidence of water. And *Spirit* is a little more hardworking, a little more hard-nosed. She went all that way to find water and she got there and there's no water, and she could have given up at that point, but she's not the kind of rover to go three hundred million miles and then give up, so . . . she gets to the Columbia Hills [which are] the size of the Statue of Liberty. . . . [S]he's only meant to be on flat terrain, and she manages to figure out how to climb this hill and along the way finds the evidence she looked for.[57]

Mark told another version of this story to Adam at the wake, punctuating each time *Spirit* could have given up in the face of adversity (but didn't) with the phrase "she's not that kind of rover!" Notably, he described *Spirit*'s apparent tenaciousness not as a quality of the team (who also didn't give up at each obstacle), but rather as a question of the rover's own personality as the tenacious "little robot that could." As his story grew to a climax his colleagues gathered around, whooping and cheering. When he finished, he challenged his friend, "If you can listen to that whole story and you can look me in the eye and say she doesn't have a personality, then *you* are the robot!"[58]

Everyone laughed. But Adam responded that for him the rovers didn't have a personality. They were always a "we," not a "she," because that emphasized the importance of collective responsibility:

> To me, I'm this human being responsible for this robot. The robot does what we tell it to do. If it doesn't do what we'd like it to, it's our fault. And so what helps me keep it rational is that *we* as humans are totally responsible for this

rover. There's no personality here. . . . There's no need to imagine a personality. . . . I'm the first one to fall for [the nautical convention explanation]: I called my rowboat a she growing up. But because I'm responsible for [commanding the rover], I feel obligated not to do it.[59]

In this moment, a conflicting form of talk surfaces alongside the members' use of "we" for the rover: the pronoun "she." Certainly an entire paper might be written about use of the gendered pronoun, which team members frequently ascribe to "nautical convention" but also explain as related to the rover's "cuteness" or "gracefulness" and to their attentive care for the robot.[60] However, it is also useful to note how the alternation between the first-person plural and third-person singular denotes a tension inherent to human-robot interactions. That is, the use of "we" versus "she" is team members' way of articulating the messy question of human-robot agency.

As I have described, "we" is by far the most prevalent pronoun in the context of strategic and tactical planning that engages the whole team. Adam's response expresses the "we" I have described in this book: producing a legible, trafficable Mars is bound up in teammates' practices that produce a collectivist orientation. The pronoun "she," on the other hand, is most frequently associated with accounts that describe the rovers' activities or their biographies. This was especially clear when the Principal Investigator jumped into Mark and Adam's conversation at the Athenaeum, explaining the genesis of the pronoun "she":

When *Spirit* first became a "she" for me was way, way before we got to Mars. Part of the reason I view them as having personalities is, if you were there when they were *born*, they went from being inanimate hunks of metal to things that actually *did something*. And moved, and behaved, and took on personalities. And I was never able to let go of that because it was as you watched them first . . . come to life for the first time, that's when I first developed that. . . . So that's where I get it from, not because of anything that happened since these vehicles got to Mars, but from seeing these things that were once sketches on our whiteboards turn into hunks of metal and then turn into living things.[61]

Note how, in the PI's account, talk about personality that deserves the pronoun "she" is associated with the notion of agency: the rover is, and always has been, something that actually *does something*. *Spirit* and *Opportunity* bring their own personalities to the experience of exploring Mars. When the rovers appear to behave as if of their own volition, when they act "temperamental," when they crush the wrong rock, when the unexpected happens, the narrative enrolls the rovers as an active part of the mission story. They become additional members of the team with a "personality."

Scholars Lucy Suchman and Morana Alač have both described how robotic agency is constructed at the interface between the human and machine. As humans interact with a robot, they assign it varying types and degrees of embodied agency, limitations, and possibilities, even despite its programming.[62] On the Rover mission, the boundaries between human and machine shift in members' accounts. As I have described, the use of "we" performs important work for the team in producing the members' social order and ways of seeing and drawing Mars. But it fails to account for moments when the robots behave unexpectedly, or for their own trajectories on Mars. Team members therefore engage in a kind of *agential gerrymandering*, linguistically drawing and redrawing the boundaries between human and machine.[63] On the one hand, the machines are an expression of the team's actions and interactions; on the other hand, they have personalities and agency all their own. The team signals this tension by alternating between "we" and "she."

A scientist on the mission once described this issue to me as one that simply revealed the paucity of the English language for describing agency in the first place. A member of the Miami tribe, he once faced the problem of how to translate "Mars rovers" into Algonquian during his public outreach initiatives with Native American communities. Algonquian words possess a fundamental linguistic distinction between things that have *animacy* and things that do not. On consulting with tribal elders, he chose to translate the rovers as having animacy, which he described as an inherent "life force": "It [animacy] essentially is an extension of us. Other *things* don't have it. Cars don't have it, trains don't. It's not a possessive language, it denotes what something *does*, not what it *is*."[64]

We might then observe the use of the pronoun "she" and discussion of personality as members' forms of talk that signal the rovers' animacy in the context of the mission. For Mark, the rovers' personalities are revealed through what they have endured on Mars; for the PI, the rovers have personalities because of how they have moved and behaved since they were born; while for Adam, the rover acts only because he and his colleagues tell it to and don't "mess up." This does not signal a fracture in the team's discourse. While "she" is more frequently used in discussion with the public or with outsiders and "we" is used in planning, the "she" and the "we" are not contradictory. Even Mark conceded to Adam that he upheld both pronouns at the same time:

> I never lose sight of the fact that there's a "we," and also there's a "she," right? I have both of those perspectives at the same time. I don't lose one just because I take on the other. . . . There's a duality in my own attitude toward *Spirit* where I see her as a "she," but I never forget that there's also this "we," and *we* have to do our jobs, and *we* have to do our jobs absolutely right and perfectly, or *she* doesn't do her job properly. But also there's two levels of it, right? There's the "we," and then there's the "us." And the *us* is who *Spirit* really is, right? That's the *she*.[65]

Amid this plethora of pronouns, the "we" implies a collective imperative alongside the agential, the animate "she." Yet even the source of *Spirit*'s apparent agential "personality" is the collective: "us."[66] At a distance of millions of miles, mediated by images, visualization software, and forms of talk, members' shifting and dual sense of robotic agency is produced at the interface of the human and the machine.

Constraints and "Lookiloo"
The Limits of Interpretation

Mars Rover scientist Sam still recalls the day in 1985 when he received his subscription copy of the *Whole Earth Review*. The front cover featured a picture of San Francisco swarmed by flying saucers. "The headline read, 'The end of photography as evidence of anything,'" he laughed. "It was all about, of course, this new application that a company called Adobe was developing called Photoshop, and they illustrated how fast and trivially easy it was to make pictures of anything . . . [such that] unless someone handed you really a negative of something you shouldn't trust it [photography] any longer."[1] Sam found the *Whole Earth*'s headline somewhat ironic because digital photography and other computational remote sensing tools had, he believed, transformed his own field from one of speculation into "science." Before digital photographs, Sam explained, planetary geologists engaged in what he pejoratively termed "lookiloo" analysis: "looking at pictures and making up stories." At the time, the "pictures" were usually orbital images taken by vidicon tube cameras onboard *Viking* or *Mariner*, the techniques of "making up" involved tracing and measuring, and the "stories" were assumptions about the geological processes at play on the surface of the planet. "People got

191

whole papers published this way," Sam laments. He contrasted this work to the Galileo mission to Jupiter and the Mars orbiter missions of 1996, which sent back digital images taken by CCD cameras that could be quantified and correlated with topographical data acquired from an onboard laser altimeter. According to Sam, this presented the planetary science community with reams of new and trustworthy information that had a whole new status. The entirety of what was known about Mars was suddenly up for grabs. As Sam delicately put it, "What we learned from Mars in the nineties is, we were full of shit."[2]

Planetary geologists like Sam regularly credit digital photography, along with associated techniques and instruments such as laser altimetry and spectroscopy, as nothing short of revolutionary in terms of their understanding of distant worlds. But Sam's description of "lookiloo" analysis highlights a tension inherent in working with digital image data. If images can be manipulated at will, what is to stop them from being "evidence of anything"? Image manipulation of the type described in previous chapters is a central part of scientific work with visual data, but this very malleability leaves such images open to suspicion—recall here the exasperated team member, faced with so much visual interpretation, requesting an image marked up to show "what's real." Scientists' distance from their field site also leaves such manipulation open to question. With all the myriad possibilities for purposeful visual construal, can digital images be drawn as anything at all?

This chapter examines the practices and accounts that members of the Rover mission appeal to in order to support their image work as trustworthy accounts of Mars. Much like the human and machine calibration work discussed in chapter 2, these practices maintain a sense of trust in images despite their repeated digital manipulation. Attention to how and why images cannot, in fact, be drawn as anything at all reveals the group policing of members' interpretative work—the implicit and explicit moral codes that govern the trustworthy production of both observations and individuals.

Central to this story is what the scientists I studied called "constraints": practices, actor's accounts, and activities that are performed and invoked to impose limits on visual interpretation.[3] The scientists I observed frequently appealed to the notion of constraint to describe several related practices in their digital image work, while simultaneously addressing the status of those images as they underwent interpretative manipulation. In the face of ambiguous images, "coloring according to your hypothesis," and a range of *drawing as* practices that variously reveal and conceal different aspects of Mars, Rover team members invoke constraints to police the fine line between what they believe to be scientific image manipulation and what they would accuse of being "lookiloo" analysis. Although this chapter will make no attempt to pronounce on whether such practices guarantee reliable knowledge, examining constraints as an actor's category reveals how scientists on the mission demarcate different modes of visual practice and responsibility within their community.[4]

Because constraints in their various forms function as the "thou shalt nots" of digital image manipulation, they indicate the proper way to make knowledge using digital materials. Thus, talk about and work with constraints is related to anxieties about the nature of digital knowledge production. But attention to constraints as an actor's category also reveals the continued importance of community-shared values of self-conduct in the production of trustworthy scientific knowledge.[5] Central here are members' actions and accounts that fellow team members recognize as providing suitably constrained interpretations of visual data. Below I will discuss three examples: mathematical rigor, combining datasets, and Mars analog work; I will then give an example of how the three are brought into conversation, each constraining the others. Since these practices constrain not only interpretation but the scientist as well, they expose the community's concerns about the reliability and limits of scientific interpretation based on data from Mars.

Mathematical Constraints

Ross's decorrelation stretches are legendary among the Rover scientists. No other images on the mission so vividly recall the palette of Andy Warhol, and Ross's images can make even the slightest difference between units or soils "pop out." When they heard I was studying images, many mission scientists helpfully suggested I meet with Ross to learn how he produces his unique and admirable images. I therefore set up a visit and planned on spending a few days at his office in a lively university town to learn more about his image work. But when I arrived and asked him to demonstrate his technique, Ross was perplexed. He started his image-processing program on his computer, loaded a set of images taken by the Pancam, and said, "I just push this button." When he clicked on a built-in function, the Pancam image turned into a brilliant decorrelation stretch with the colors that unmistakably marked it as one of his works (fig. 7.1). Although at first I was somewhat disappointed that this unique production could be ascribed to a built-in software function, I quickly learned that what was important for Ross was that the button initiated a coded script that applied a precise mathematical formula to the images he had selected. Thus these images were not transformed

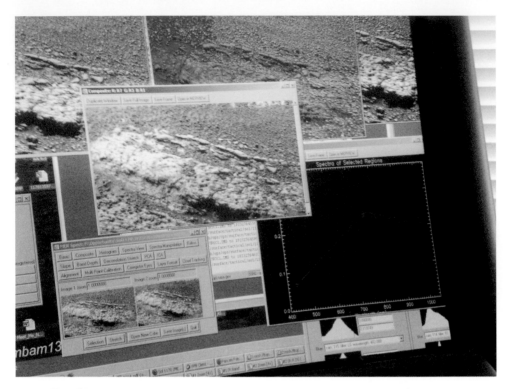

Figure 7.1. Ross uses built-in software macros to compose his decorrelation stretches of Pancam images of Mars. Author's photo.

or interpreted willy-nilly but maintained a persistent underlying mathematical integrity with respect to the original dataset.

Many scientific digital image-processing practices invoke mathematical expressions related to geometry, functions, integers, and operations. Pixels are added and subtracted, multiplied or divided, and may also be subject to complex equations or derivations. Ross frequently plots pixel values in three-dimensional graphic space, looking for clusters of dots on the graph that might identify mineralogical commonalities between parts of the visible image. These graphs may themselves be subject to further functions. Ross impressed on me the importance of computing eigenvalues, features of matrix algebra that determine the vector relation between pixel values as they are plotted and manipulated in multidimensional space (fig. 7.2).

The appeal to the mathematics of the digital image is not specific to planetary science. Sociologist of science Anne Beaulieu has described how fMRI

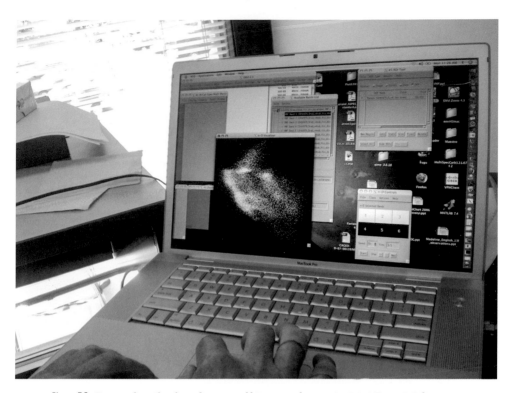

Figure 7.2. Ross works with a three dimensional histogram plot, rotating it in 3-D space and computing vector relations between pixel values. Author's photo.

researchers and clinicians practice a kind of "iconoclasm" in their approach to the "pretty pictures" produced by their machines, preferring to describe them as representations of statistical datasets. In Beaulieu's analysis, clinicians and scientists adopt different stances toward the pictorial and numerical aspects of their data, speaking to a divide in status and disciplinary heritage.[6] Historian of science Peter Galison has also described the productive clash between the numerical tradition and the visual tradition in particle physics, as different technical and visual cultures produced interpretations of bubble chamber images.[7] Unlike the scientists that Galison and Beaulieu describe, however, Rover scientists do not privilege one form of visualization over another in determining the status of analysis. The numerical and the pictorial stand instead as different aspects of the same dataset, each brought into relief through *drawing as* practices. Different visions of the same dataset are frequently brought into conversation with each other, evaluated for what each one shows best, and limited by what they do not show. In the context of scientific work, however, the appropriate invocation of each aspect is critical in presenting and supporting claims. Scientists across the mission repeatedly invoked the numerical as a constraint on pictorial interpretation through two mathematical techniques: appeal to replication and practices of combination.

A year later I was reminded of my visit with Ross when I was writing a paper that included one of his characteristic images but realized I did not have his permission to publish it. When I contacted another member of the team who frequently works with Pancam images to ask how to get permission, I was instructed to just credit it "the usual way," with the mission's institutional tagline for all Pancam images. I insisted that this was very clearly Ross's image and recalled my conversation with him about the values of artistic production in science and even the possibilities for gallery exhibitions of some of his more striking pictures. But the other scientist countered that "anybody could have made that image" and then suggested that if I was still concerned about it, he could recreate the image on the spot and give me permission to publish that instead. I must have seemed taken aback by this offer, so he elaborated. What made Ross's work scientific was precisely this ability to recreate the image. Because his image was a combination of particular filters governed by a mathematical formula, it was and should be replicable. Indeed, his images' very status "as evidence of anything" depended on their ability to be precisely recreated at will by any interested scientist.[8]

The Rover scientists' interest in replication is perhaps not surprising given the importance of replication in the experimental sciences.[9] But this does not mean that group members routinely replicate each other's work as a way of fact-checking, confirming, or undermining another's experiment. Instead, replication is invoked as a constraint on interpretation. Manipulated images can be replicated, Rover scientists explain, because they were created in the first place by a mathematical expression: a function applied to a range of numerical pixel values. If they cannot be replicated, that is because the underlying mathematics has been tampered with in an unpredictable way. Thus the mathematical constrains the visual but does not preclude, supersede, or mingle with it. The numerical appeal restricts scientists to those types of image manipulations that can be replicated, mathematically described, or generated. So this constraint on image interpretation as it is enacted in practices of image work not only disciplines the image into a trustworthy document as it is *drawn as* to incorporate an interpretative move: it also disciplines image processors as they draw.

Another constraining practice is one of combining datasets from different instruments, spacecraft, or scales. Rover scientists call this coregistration. The scientists I interviewed explained coregistration as "a desire maybe to see different things at the same time," as visualizing context, or as "using data to give you confidence in other types of data." While coregistration is a kind of *drawing as* practice through the digital combination of datasets, the commitment behind the practice is a belief that the datasets would not align were there no naturally existing mathematical correspondence between them. As one scientist explained it, "You can't make a mosaic unless all your pieces are from the same puzzle." Data manipulation must not alter or otherwise upset this natural correspondence or the datasets will not align. This practice therefore reflects and supports a correspondence theory of representation, while at the same time constraining interpretations of the pictured object.

It is especially common to coregister spectral and pictorial datasets, such as readings from the rovers' MiniTES and Pancams, or from several generations of orbital spectrometers like THEMIS, OMEGA, or TES or the cameras HiRISE, HRSC, or MOC. Indeed, when I asked Ross to explain the difference between a spectrometer and a camera he explained that the labels were often interchangeable. As he put it, cameras like HiRISE and Pancam provide spectral datasets; they just do so in the visible range of light, unlike TES or MiniTES, which take readings through infrared and thermal sensor bands. The primary difference between them is a trade-off between spatial versus spectral resolution. Most scientists choose a camera filter that can provide a high-resolution base map on which to coregister other observations, so that at least one dataset in the combination provides the essential spatial, locational, or geomorphic data while the other provides spectral data. For example, the Pancam, HiRISE, or MOC can provide a picture of visible features, which can then be coregistered with invisible features provided by other instruments, such as spectral bands in the infrared.[10] Coregistering data in this way constrains interpretations. Although alone each individual dataset could support a variety of interpretations, taken together they may point to only one or two shared hypotheses.

Let's return to Ross's desk for an example. When I visited him, Ross was trying to make a case for a landing site for the next Mars rover, *Curiosity*, and so was working with orbital imagery to show that the region he was interested in presented evidence of past water activity. To do this he needed to combine (or coregister) orbital images from HRSC, the High-Resolution Science Camera onboard Europe's Mars Express orbiter, with images from OMEGA, a spectrometer onboard the same orbiter (fig. 7.3). Or as he put it, he needed to "overlay . . . those mineralogical interpretations onto this multispectral data."[11] Ross selected these datasets because they provided coverage of the part of the planet he was interested in, and

Figure 7.3. Ross combines HRSC visual and OMEGA spectral data from the European Space Agency's Mars Express orbiter. The false color HRSC image serves as base. The OMEGA signal for the 1.9 micrometer water band display in red pixels on the image overlay. Author's photo.

because he wanted to be able to correlate the detection of water (a spectral signature) with its spatial (geographical) location on the planet. He calls the two views of the planet "complementary," since one has "much higher spatial resolution but lower spectral resolution" but the other has "higher spectral resolution but lower spatial resolution." Ross used ENVI, a commercial image-processing software suite, to combine these two sets of data. He used the latitude and longitude measurements encoded in the image files to create a correspondence between the two datasets,[12] then projected both datasets onto the same map of Mars, one on top of the other. In the resulting image, the camera provided a sense of context, the pictorial features of the terrain, while the spectral data showed up in a variety of colors painted across the scene. Looking at the data in this way, Ross noticed that the areas with a water signature somewhat lined up along topographical features visible from space: the flatlands around the rim of a crater. This led him not only to identify where those water features were located on Mars, but also to make a claim about where they came from and their relationship to the nearby crater.[13]

The combination of both qualitative and quantitative datasets is an important analytical and rhetorical resource. Sam spoke plainly of the importance of this dual appeal as evidence in supporting hypotheses. He recalled a situation in which climate modelers, who worked with simulation software,[14] were at odds with geologists working with *Viking* image data over how to model the environment of early Mars: "People were so sure that precipitation and run-off weren't happening [on Mars] because [the atmospheric modelers] couldn't make it work. Geologists could only offer a qualitative explanation, [but the modelers would say] . . . 'you can't show us any evidence for that, you just sort of handwave at your pictures. . . .' You had all these climate modelers armed with numbers and physics against all these geologists armed with only our pictures."[15]

The hypothesis of gullies and runoff created by precipitation looked suspiciously like "lookiloo" at first, and the mathematical model was believed over the visual. But once quantitative topographical data from the Mars Global Surveyor satellite was coregistered with the orbital images, the geologists had the quantitative measurements—as Sam put it, the "evidence"—to back up their visual interpretation. At this point they could charge the atmospheric scientists with believing too much in their computer model and not tempering this approach with experience. "We went [back] to the climate modelers and said, 'Now it's your problem,'" Sam recalls.

This talk and activity with digital images reveals a belief in an underlying mathematical rigor that lends trustworthiness and evidential status to the resulting images. But not only do these constrained practices of digital processing discipline the image into a trustworthy document, the activity disciplines the image processor too. Katie, a graduate student on the mission, explained this to me as she compared orbital and ground-based images of Mars taken by the rovers and the MRO spacecraft's spectrometers. When I asked if she couldn't just combine the two datasets visually, eyeballing the result, she emphatically objected:

> You really have to [do the comparison] mathematically, and it's much more scientifically rigorous to do it that way anyway . . . because if you look at two images and you say, oh these two look the same, you can't really, it's hard to get that published. . . . Scientists are like, "That's subjective! They might look the same to you but they might look completely different to somebody else!" Science has to be backed up by . . . statistically significant results in order to make sure that you're making the right interpretations.[16]

Note how Katie associates the mathematical approach to her coregistration problem with being "scientifically rigorous." This rigor is also associated with the difficulties of publication and the complexities of subjective observation. The appeal to mathematics, to rigor and statistics, supports her analysis as "the right in-terpretations." But Katie's language of subjectivity, publication, peer evaluation, and rigor also invokes the scientific person, the practices and identity of the responsible scientist whose interpretations can be trusted and whom she hopes someday to become. Adopting the mathematical constraint on digital manipulation helps her shape her own identity as a responsible observer, image processor, and scientist.

Fieldwork as Constraint

The discussion above places much emphasis on the quantitative over the quali-tative. But an equally important constraint to visual manipulation is the appeal to experience and judgment. Unlike judgment based on digital visual experi-ence, as in the case of calibration, here judgment is gained from experience in Marslike environments on Earth. The remote nature of the planetary scientists' work sites produces an analogical step in generating experience to justify an in-terpretative claim about Mars. Even as Rover scientists describe wanting to "get their boots dirty" by stepping into their robots' tracks on Mars, they make con-sistent use of field or laboratory studies on Earth that they consider analogous to the sites they are examining on Mars. In a curious juxtaposition to the digital nature of rover image data, these Earthbound field sites are called Mars analogs.

Fieldwork has an intriguing status in planetary science. Earth-based fieldwork has been an essential part of any planetary geologist's training from the early his-tory of the field. Historian of science Maria Lane's study of late nineteenth- and early twentieth-century Martian observations notes the importance of going into the field; at the time, this consisted of parties of astronomers' traveling to remote sites where they could better view Mars through their telescopes with-out the variations of atmospheric opacity and light contamination from cities.[17] Later in the twentieth century, planetary geologists were required to practice field techniques such as geological mapping on Earth before they could apply their skills to extraterrestrial sites. Even today, a planetary scientist's training involves working with orbital images of Earth, drawing on them to transform them into maps identifying particular types of terrain, stratigraphic layers, or mineral deposits (fig. 7.4), then taking these orbital images into the field, walk-ing carefully around the area on Earth to better understand how what is on the ground is seen from space. As Stefan Helmreich has claimed in his analysis of astrobiology, through such practices the Earth becomes something more than itself, a representative of planets as a general category, and a laboratory through which scientists explore what planetary environments might be like.[18]

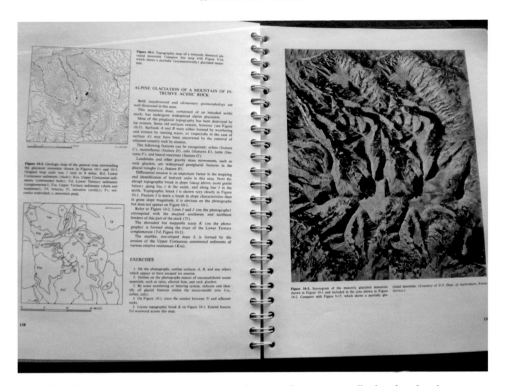

Figure 7.4. A geological mapping exercise using drawing techniques to visually identify and mark stereo aerial photographs of a field site. From Miller and Miller, *Photogeology*, 138–39.

Rover scientists employ particular parts of the Earth as Mars analog sites, and they both refer to and visit them regularly. These include Río Tinto in Spain for its high-iron and highly acidic groundwater; elevated, ultradry deserts in South America; and research stations in Antarctica or the high Arctic. These analog sites are not meant to fully replicate or simulate a Martian environment; instead, the language of analogy invokes another constraint on the interpretation of digital image data. For example, a meteorite expert used "samples we have in our labs" to develop a hypothesis about how particular meteorites on Earth— and by extension, one under discussion on Mars—undergo changes when exposed to water.[19] Another scientist pointed to the distribution of meteorites on Antarctica as a case study that would enable the scientists to "confirm or refute [an existing] hypothesis . . . [and] assist in confirming the meteoritic character of [an individual rock under study]."[20] A senior scientist and respected planetary mapper insisted to his colleagues at a Team Meeting, gesturing to the ripples in the Martian terrain visible in an orbital image of Mars, "Of course, if we'd seen this image on Earth there'd be no question that this would be formed by water."[21] Such statements are commonplace across the mission and the planetary science community more generally, where scientists propose interpretations of terrestrial materials to suggest, support, or challenge various interpretations under consideration about Mars.

Collecting rock samples on Earth is another important aspect of rover work on Mars. When I visited the meteorite laboratory at the Smithsonian Natural History Museum in Washington, DC, the Rover team scientist there brought out a few small samples of meteorites believed to be Martian rocks, including the famous ALH84001.[22] But even without access to rocks from Mars (which may be contaminated from their forced ejection, journey to Earth, and exposure to the Earth's atmosphere and biosphere), scientists turn to Earth-collected samples of the minerals they see on Mars. When I visited them to observe their image-processing work, both Ben and Ross were quick to pull out their field samples of terrestrial rocks that betrayed some quality, whether in mineralogy or in texture, similar to the ones they were visually construing on Mars.

Scientists also frequently appeal to terrestrial laboratory conditions, often altered to approximate some aspect of Martian conditions. Many team scientists maintain active laboratories of all shapes and sizes that sport equipment from spectrometers to wet labs, pressure chambers to chemical apparatus, and even sandboxes with simulated Martian soil that churns under surplus rover wheels. Susan, for example, has two laboratories, one in which she builds new spectrometers to test and eventually propose to upcoming missions, the other in which she performs chemical experiments to approximate weathering conditions on Mars (fig. 7.5). Susan used this laboratory to constrain her interpretations of Tyrone's changing soil spectra or, as she put it, "to be sure this change is real": "We need to be sure this change is real, so I checked several factors. . . . One possible change could be the dehydration of hydrous salts. . . . I did an experiment starting with seven water ferric sulfate."[23]

The experimental results suggested that ferric sulfate could change, and they determined under which conditions the results she saw in the Pancam spectra might be effected. Describing the experiment to me in a later interview, Susan called it "observation and laboratory experiment put together, and some common knowledge."[24]

When another scientist I visited opened drawer upon drawer of carefully collected samples from field sites on Earth, explaining to me that he liked to "get samples, get things in my hands" (fig. 7.6), I asked if this made the kind of science

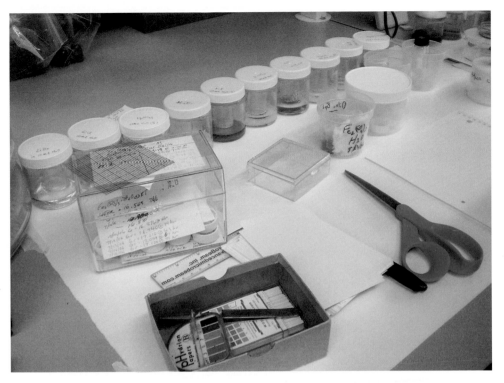

Figure 7.5. Susan's experimental work with ferric oxides to constrain her visual interpretations of Tyrone. Author's photo.

he did on Mars using rover data seem somehow poorer in comparison. He said no: "Even if I can't get samples [from Mars] in my hands, they've [the rovers] done a good job."[25] Again, neither the fieldwork nor the digital image work is considered superior, but each is construed as an aspect of the same kind of object, the same work. They are also invoked as checks each on each other. While each aspect may reveal different features of the imaged phenomenon, they may not reveal inconsistent or incompatible features. These different aspects and elements of experience must be brought together to produce trustworthy representations. As another team member put it, "Bright kids can make computers sing and dance—now they have much better technical skills—but what they don't have is twenty-five years of being in the field."[26] The appeal to field experience doesn't trump but rather constrains digital interpretive work with images alone.

For example, at the Team Meeting in July 2007, Susan was again on the agenda to discuss preliminary Pancam results on a target other than Tyrone. At

Figure 7.6. Drawers filled with samples collected in the field, approximating Martian geological phenomena. Author's photo.

her presentation a few months before, she had shown her decorrelation stretches and histograms of Tyrone, appealing to the pictorial and the numerical sides of the image; she had also presented results from her experiment with ferric sulfates to constrain her hypothesis. But when Susan presented a new study of a similar feature and this time presented only Pancam spectral work, her teammates met her conclusions with skepticism. In the question period following her talk, Ross accused Susan of "overinterpreting," suggesting that some of her data were "obviously dust affected." His challenge was rooted in his sense of the field environment: as he saw it, Susan was interpreting her spectra without considering the field context in which they were embedded, such as which minerals would likely be seen together and how much dust contamination such an area would imply. Fresh out of his own laboratory, Nick asked Susan, "Are there any lab data . . . that support or sort of suggest what that feature is attributable to?"[27] Such questions, amid others, revealed the group's discomfort with appealing to the visual and mathematical side of the image alone without considering "experience" as a constraint on interpretation.

This exchange exemplifies the importance of counterbalancing an appeal to both kinds of constraints, the digital and the analog. Leaning too heavily on the computational side can involve the scientist in circular logic with no window onto reality. But too much reliance on visual interpretation from field experience on Earth can be challenged if there are no digital data to support such interpretation. Thus multiple kinds of constraints are called on in practice, such that each type of knowledge production constrains the other. As one of Sam's students put it, describing his resistance to presenting data from his simulated model of Mars as an explanation for the Martian environment, "Computers only do what you tell them to do."[28]

As the Rover scientists move back and forth between their digital experiences of Mars and their physical experiments on and interactions with Earth, this complicates the border between the lab and the field and allows for some mobility of techniques and interpretative frameworks between the two.[29] Scientists are simultaneously in the field and in the lab, whether they are at their desks or immersed in an analog environment. When Susan is in her lab or Nick is at his spectrometer, they are by simulation "in the field" on Mars; when Ben is at his desk manipulating an image so as to see something new, he is also "in the field" on Mars. Similarly, when Gwen is "in the field" at Lake Tahoe or Yellowstone, she is also "in the lab" in the sense that the environment she seeks to understand (Mars) is only partially simulated (on Earth).

Even outside the lab or the field, Rover scientists are busily engaged in blurring the boundaries between the two, or between Earth and Mars. After the discovery of the blueberries on Meridiani, graduate students on the mission celebrated by flooding their adviser's office with hundreds of spherical hematite concretions collected in Utah, effectively turning the area into "the field" on Mars. The Principal Investigator regularly dresses in jeans, a plaid shirt, and cowboy boots, often saying aloud that he wishes he could just pick up his geologist's hammer and walk out into the field, right into the images of Mars on his screen. These practices are analog work too, building a repertoire for Mars through experience on Earth. But they also constitute visual work, since such practices produce the locally approved experience and judgment that Rover scientists call on to constrain their interpretive work with images of Mars.

Constraints in Action

Constraints not only affect interpretations of singular images or instances of image processing, they are equally important for building chains of association between images and interpretations, to produce scientific discoveries. Bruno Latour has described this process of chaining inscriptions together in his study of a soil science field site in the Amazon, while sociologist of science Trevor Pinch describes not only chaining together inscriptions but also producing "a chain of interpretation." In his study of solar neutrino detection, Pinch describes how scientists, when confronted with the same graph, stop saying they see splodges on a graph and start saying they see argon atoms or, eventually, neutrinos. Each move along this chain of interpretation requires a tremendous amount of work to establish that this splodge may be *seen* not just *as* evidence of a neutrino, but simply *seen as* a neutrino. Pinch frames this as a question of varying degrees of externality in observational reports. Certainly we might say that seeing a splodge on a graph as a neutrino inserts an interpretative intervention. But as Pinch points out, each move along this chain is more precarious than the last. At each step, alternative interpretations are closed down, producing a possible observation that has been interpreted, yet whose interpretation may yet be underdetermined, standing on shaky ground.[30]

Invoking constraints on interpretation allows Rover scientists to make what they consider valid moves along a chain of interpretation and to incorporate greater degrees of externality in their observations. Recall how Susan used her laboratory data to constrain her interpretation of the Pancam spectra and could therefore move from an observational report about seeing changing histograms to seeing the dehydration of hydrous salts. Thus planetary scientists may appeal to the trustworthiness of their observational reports as, or even because, they actively manipulate the very data that constitute their observations of Mars. But constraining hypotheses constrains the scientist as well. After all, if hypotheses can be validated only by narrowing down interpretative flexibility and delimiting possible interpretations of data, but the data that return from the rovers is always to a certain degree underdetermined and require manipulation in its very analysis, then scientists can get trapped between evidential contexts in their observational reports.

A powerful example of such constraints in action is the case of silica sinters at Home Plate. After the discovery of the salty soils at Tyrone, *Spirit* returned to Home Plate, taking pictures of every unearthed patch of white soil along the way. But examining these pictures, Rover scientists began to notice other features as well: oddly textured small rocks that the team initially called cobbles. The cameras and spectrometers on the rover identified silica, but it was unclear whether the silica was a coating or a component of the rock. If it was a component, one could say that the rock was built up by silica deposited within a hot spring environment. If it was a coating, it might be a remnant of some transformation to the rock's surface effected by steam or some other hydrothermal system. The team commanded *Spirit* to approach examples of these cobbles for a closer look: in chapter 5 we saw Susan, Jane, and Alexa negotiate which one to image.

At a Team Meeting a week after the targets were selected, Nick, a MiniTES specialist, reported surprising results. The cobble targets in question were registering over 90 percent silica in their composition. He described two possible hypotheses for the rocks' formation—one as a deposit of pure silica, the other as a coating—claiming that they were likely "distinguishable from one another as a function of silica content as a function of depth." But another scientist interjected that that was not so: "You can see in Hawaii, for example, there are coatings of opal and silica that are sitting on top of [the grains that make up the rock]." That is, the grains that make up the rock, not just the exterior of the rock, could themselves be coated with silica, and this would indicate yet another geological process. Thus the depth of silica presence within the rock could not constrain a hypothesis about how that silica got there. Another scientist agreed, "I have exactly that from Hawaii . . . where I scooped up sand and [examined it] under a microscope," to which another assented, "the Hawaiian silicon coating is a classic." Gwen, meanwhile, put up PowerPoint images of silica systems in Wyoming and Nevada, promising to conduct further fieldwork studies in Tahoe to investigate.[31] This talk about Hawaii and Nevada was actually talk about Mars, an appeal to analog environments as a first level of constraints for what the observations could mean.

Ultimately, one of the scientists in the room bemoaned the fact that "the presence of silica does not constrain a depositional environment. . . . Silica is just too complicated, too ubiquitous to nail it down." The conversation then centered on which features the science team could look for on Mars that *would* "constrain a depositional environment." The group generated a list of observations the team could accomplish on Mars and on Earth to constrain their interpretations, and the list was entrusted to an LTP Lead to implement. At follow-up End of Sol meetings, a group of the team's geochemists presented results from laboratory work, while geologists produced samples and images from terrestrial sites.

The outcome of this extended exchange was a decision to use *Spirit*'s wheel to crush one of the cobbles, then take a picture of its interior structure with the Microscopic Imager and use the spectroscopy suite to produce coregistered observations of its chemistry and the chemistry of nearby soils. A target was selected based on its high silica content visible in MiniTES spectra.[32] When the images and readings returned from Mars, the excitement mounted. The MI pictures displayed a texture that the geomorphologists recognized as a sinter. On Earth, sinters are made from deposits in hydrothermal spring systems, like those at Yellowstone. The spectroscopists, for their part, explained that the rock's composition could be identified as a kind of opaline quartz. On Earth, opaline sinters can be formed by microbial processes in hot springs. Listening to the End of Sol meeting that afternoon, I noted that the energy on the line was electric. As the scientists cautiously traded theories about how such sinters could form on Mars, my field notes recorded the first use of the word "biology" since the start of my ethnography over a year earlier.[33]

At this stage, several rover observations and several constraints were already in play. MI images had been analyzed for their geometries and textures. MiniTES spectra had been analyzed for relevant peaks and dips in the instrument's data. Samples had been collected and analyzed from Yellowstone and Hawaii, and chemicals had been collected in a laboratory. The rover had been commanded to move, then to take more photographs and readings, participating in the story. But how to move from saying "We see a cobble," "We see texture," and "We see peaks on a graph" to saying "We see a silica sinter," "We see evidence of past hydrothermal activity," or even "We see evidence of past life on Mars"?

Bringing these different interpretations together into a chain of reasoning requires satisfying numerous actors' constraints to tightly couple the relationship between an evidential report and a change in interpretation. But as the science

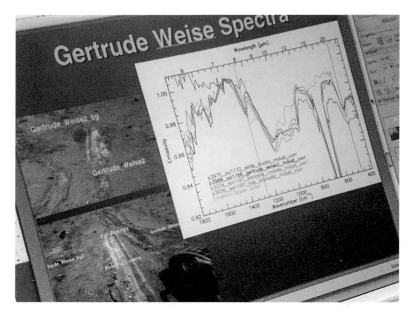

Figure 7.7. The spectrum of Gertrude Weise, produced and interpreted by Nick. Note the use of rover images and annotations to mark the location of each spectrum's acquisition. Author's photo.

team members attempted to rigorously constrain their hypotheses, they were restricted from making just any analytical, digital, or analogical move. This restriction was due to the wider social implications of presenting underdetermined claims to the broader community, which could accuse such claims of being unfounded or methodologically invalid. This required scientists, uncomfortably, to confront the community's anxiety about the status of their knowledge claims. Nick was caught in this very bind.

When I visited Nick, he was beginning a set of laborious analog observations that had him both excited and cautious. Reviewing the spectra that returned from the patch of soil named Gertrude Weise, Nick had noticed a small "bump" in the MiniTES spectrum around the eight micron region (fig. 7.7), which he had described on an End of Sol teleconference line a few weeks earlier. He pointed it out to me in the middle of the squiggly line, directing my attention with gestures at the screen, "It starts to have this feature here [points to a part of the spectrum]. . . . [I]t's what a spectroscopist would call a shoulder. . . . [B]y the time you go to these siliceous sinters with their very distinctive texture, what I'm discovering is that that shoulder turns into a fully resolved minimum, an absorption minimum there [points to the eight micron spike in the spectrum]."[34]

Working with MiniTES observations, much like Pancam readings, requires considerable skill and a combination of *drawing as* and *seeing as* techniques that betray Nick's professional vision as a spectroscopist. Even identifying a feature amid the MiniTES graph betrays mastery of a technique. When I locate another spike in the spectrum, he dismisses it: "That's in the lab, that's an artifact that wasn't removed, so that's a total garbage thing."

In his analysis of the MiniTES graph, Nick can appeal to both the pictorial and the numerical to constrain his interpretations. The graph is composed of numerous discrete points that can be mathematically transformed, added, subtracted, and combined. The combination of different elemental materials will change the shape of the graph, so spectroscopists like Nick frequently engage in spectral deconvolution: computationally removing suspected elements and minerals from the graph by subtracting their values from the graph to see what remains, or even adding together mineralogical compositional elements to see if the computerized graph comes close to the observed one. But understanding what the spectroscopists refer to as "squiggly lines" also requires constraints arising from considerable judgment and experience, both with the instrument and its results and with field samples and geological context.

Silica, for example, has many forms corresponding to different formational processes. Each produces a different spectral graph. Nick therefore used a combination of his computational resources and his experience with spectrometers to identify the kind of silica that would produce the eight micron feature. At first he thought the peak was due to quartz, but when he loaded up the spectrum for quartz from his computational spectral library, the visual comparison revealed a difference. He loaded the two spectra side by side, that of quartz on Earth and that from Gertrude Weise on Mars, and talked through what he saw as the similarities and differences between them: "This peak [on the quartz spectrum] is what I was thinking I was seeing in this spectrum [Gertrude Weise], so I thought this black peak [quartz] is this purple peak [Gertrude Weise], but it's shifted so there's no way that it's due to quartz."

Constraining his interpretation required eliminating quartz from the equation. But when Nick loaded the lab measurement of classic amorphous silica, he detected a match between the location and pronounced bump on both graphs. Describing the connection to me, he appealed to what he knew about amorphous silica from fieldwork on Earth. "You can go to these fumerole environments in Hawaii and see this . . . effect on basalts," he explained. "[Fumerole environments] leave behind a spectrum that looks like that [Gertrude Weise]."

At this point, Nick's observational report—and achievement—had changed from a feature embedded in a squiggly line to an absorption minimum, thanks to his experience with spectral readings. It changed from there to silica with an amorphous, not a crystalline, structure, via the constraint of mathematical and experiential judgment of comparative spectra, and finally to a material "produced on Mars under fumerole conditions like those in Hawaii on Earth." This latter appeal to Earth had an important meaning. On Earth, quartz is a crystal, not formed biologically. But some types of amorphous silica (like opal) are formed through biological processes. The next step was therefore to be able to either prove or deny the claim that what was visible in the Gertrude Weise spectrum was evidence of past biological activity on Mars.

Over the following months, Nick resorted to further combinations of these techniques to constrain a hypothesis about the biological origin of the silica deposits. He calibrated the spectra from Mars with algorithms that could account for local dust, compared and computed with spectra from his spectral libraries. He coregistered MiniTES stares with Navcam and Pancam images to show exactly where the high silica readings were coming from on the surface. Nick spent hours in his laboratory with terrestrial silica samples, which he credited as "absolutely essential to understand what I'm seeing" on Mars. Consistent with the collective approach of the Rover mission, Nick did not work alone. Gwen sent him spectra from sinters she had collected in the Yellowstone area, and another scientist sent by courier boxes of silica sinters he had collected in the field. Throughout the summer of 2007, Nick sat in his laboratory and loaded the samples one by one into the MiniTES-like spectrometer that he had built as a graduate student, reporting back on his efforts every week at the End of Sol meeting, attempting to build a spectral library of different kinds of Earthbound amorphous silica deposits to compare with the Martian examples (fig. 7.8).

This laborious process did produce some of the experimental results Nick needed to add "robustness" to his interpretation of the silica sinters. In a presentation at the Team Meeting in July 2007 he brought together spectra from "Mars, Earth-Hawaii, and Earth-Yellowstone" to show a "very robust and undeniable . . . match in this eight micron feature"[35] between the Yellowstone and Martian examples of sinter deposits. At the next meeting six months later, he again went "from Mars to Earth" and showed the "very interesting, very rich spectra from Mars" alongside the ones he had generated with his spectrometer in the lab.[36] Nick declared this a case where there was a "beautiful . . . synergy" between the lab environment and the field site on Mars, strengthening his appeal to laboratory-based constraints on interpretation. At SOWG meeting after

Figure 7.8. Nick places samples in his Earthbound spectrometer to compare the readings with those of MiniTES observations of Gertrude Weise on Mars. Author's photo.

SOWG meeting, Nick gave downlink reports on completed MiniTES stares of sinterlike rocks around *Spirit* to build up a map of their location at Home Plate, and he requested further stares of other nearby objects for completeness.

Nick's attempt to properly constrain his hypotheses and avoid accusations of underdetermination required a labor-intensive and time-consuming attempt to collect and compare the spectra of silica samples, both on Earth and on Mars, to the MiniTES spectrum of Gertrude Weise. A single example of amorphous silica that was biotic in origin but did not display the eight-micron feature would falsify his hypothesis. Until that sample was found, however, the measurements continued in earnest and even lent a positive air to the interpretation. "The more of these measurements we make," said Nick, "the more difficult it is to come up with an abiotic way to make this feature, the more compelling it is." Pointing to an opaline sinter plot displayed on his screen as the rock sample sat in the spectrometer a few feet away, he explained, "This is currently the best fit to what we see on Mars, so that allows me to tell people that this opaline silica story is the best, most consistent fit to MiniTES [results]."[37]

But alongside the work of constraining his interpretation was Nick's own sense of constraint as he was caught between degrees of externality. Nick and his Rover colleagues were all too aware of the enormous implications of announcing any discovery of life on Mars.[38] As he transferred rock samples from their heated chamber to his spectrometer, Nick explained:

> I'm trying to do this myself, to be very dispassionate about it, because on the one hand it's like, shit, have we discovered life on Mars? On the other hand it's like, come on, it's not that easy. I totally subscribe to [Carl] Sagan's classic quote that extraordinary claims require extraordinary evidence. . . . This is an extraordinary case. The evidence is compelling so far, but it's not extraordinary.[39]

Nick appeals to intensive laboratory work, a sense of dispassion, and a requirement for extraordinary evidence in an attempt to either affirm or deny the next step of the interpretative chain and simultaneously demonstrate his personal, scientific moral restraint. But he also felt personally vulnerable should he attempt to make the leap. "People are gonna think you're crazy!" Nick exclaimed: he would be accused of "looking at his squiggly lines and thinking he sees bugs!"[40]

Despite these months of careful work, no discovery was made or claimed. The eventual conclusion was that the eight-micron feature was an effect of the spectrometer's viewing angle, not a property of the sample itself. Presentations of the results were therefore limited to hypotheses lower in the chain of interpretation. A paper presented at the Seventh International Conference on Mars offered the modest and reportedly "well-constrained" observation that the material in question was compared with fumerole deposits in Hawaii and that it "has a really nice match to opaline silica."[41] The Rover team subsequently published a paper about the investigation in *Science* under the reputable first authorship of the PI and with the weight of a considerable number of authors from the Rover team.[42] The paper made only the limited case that the spectra and their link to hydrothermal conditions were "important for understanding the past habitability of Mars because hydrothermal environments on Earth support thriving microbial ecosystems." Perhaps not coincidentally, however, it was printed in a special issue on microbial ecology.[43]

Conclusion

In the midst of all the fluidity, malleability, and hybridity of rover data, Rover scientists self-impose limitations on their image-processing activities to constrain their interpretations of distant planets that they have no opportunity to visit in person. Asking "Is it repeatable?" "Is it combinable with other datas-

ets?" "Can it be replicated in the laboratory?" or "Is there anything like this
on Earth?" enforces restrictions on community members' data manipulation
such that the resulting images cannot be taken as "evidence of anything." These
practices do not exist in isolation, either. While one or another scientist may be
more familiar with one practice or another, from decorrelation stretches to wet
chemistry, they believe these practices are necessarily complementary. Working exclusively on the computational side can place the scientist too far from
the Martian field site; working in the laboratory with no concept of what is being seen on Mars presents the same problem. Each practice is therefore locally
invoked as a constraint on the other. And as the different visualized aspects
are brought into conversation with each other in this collective endeavor, they
also enable and constrain movements along a chain of reasoning to develop
increasingly externalized observational reports. Attention to constraints as an
actor's category thus usefully reveals both the practices of local epistemological
work and the principles of moral engagement, especially as they relate to visual
interpretation.

However, the constraints scientists associate with interpreting their data
similarly constrain their own behavior. As scientists like Nick attempt to constrain their interpretations with enough rigor to satisfy their colleagues, they are
restricted from making particular analytical moves or pronouncements. This
restriction is bound up in the wider social implications of presenting underdetermined claims to the broader scientific community, to be sure. But it also
requires scientists to confront the community's anxiety about the status of their
knowledge claims about distant worlds. While the MiniTES spectra require
this very expert approach to interpretation involving visual, mathematical, and
experiential practices, Nick's intensive hours in the laboratory day after day, attempting to appeal to all these various constraints, revealed the reality of his
Sisyphean task.

Despite these appeals to constraints and to disciplined behavior, then, the
status of remote observations of a distant planet remains tenuous at best. Working with images or even with analog materials in an Earth-bound laboratory
can allow only for limited interpretation and limited degrees of externality in
observational reports. Rover scientists therefore recommend a substantial dose
of hubris. In Sam's words: "You need to go see how it really works. If you think
you can just look at a picture of a planet and make out its geology, then you
aren't approaching your field with sufficient *awe*. . . . What you're proposing to
do should be extremely intimidating, and you should probably accept that you'll
probably get most of it wrong, and probably get all of it wrong."[44]

"Surviving Politically" and the Martian Picturesque

A few weeks before making the decision to send *Spirit* to the northern side of Home Plate to endure its third winter, the Rover science team assembled on the teleconference line for an End of Sol meeting to review possible winter haven locations. The meeting opened with some bad news: the next-generation rover, Mars Science Laboratory (MSL; later named *Curiosity*), was over budget. Based partly on the rovers' own success, MSL had a funding level associated with the most significant of NASA projects but was falling behind its launch schedule for 2009.[1] With the NASA budget for the year already approved by Congress, no further funds were available to support MSL. The Science Mission Directorate's Associate Administrator demanded that the Mars Exploration Program manage the overruns within its branch of the organization's allocated budget. The administrator for the Mars Program was now poised to cut budgets to existing missions, including the two rovers. Relaying this news, a team member explained that the team had new imperatives to consider in surviving the Martian winter:

> What I simply wanna drive home to everybody [is that] . . . retreat is not an option for us right now. . . . The Mars Program's under

a lot of pressure these days, MSL is facing overruns; those of you involved in the MRO [know that we're facing] budget cuts across the board. . . . It's not enough that we keep the rover alive, it's more important that we keep pushing hard and getting science that is new. . . . I can say with a pretty high degree of certainty that if we were to [retreat to the north side] we would have the keys to the rover taken away from us because we're not being efficient scientifi- cally. . . . We have to push hard to the south and get as much science as we can with these vehicles . . . to survive the winter.[2]

Although driving to southern parts of Home Plate would be physically dif- ficult for *Spirit* and could possibly cause the rover's demise, driving back to the north side might also mean certain death through the denial of continued mis- sion funding. Reviewing their options, Sarah made an insightful comment: "We need to be aggressively productive during this time in order to survive both physically and politically."[3]

So far I have focused on the local, situated team context in which *drawing as* practices are embedded, describing images in interaction through digital image-processing software, with robots, and among team members. This ap- proach emphasizes the interactions through which local order is produced and maintained. But foregrounding these interactions can occur at the expense of the institutional framework in which the rovers are also embedded. Note, for example, that many of the anxieties that plagued team members like Nick in the previous chapter were due to the institutional pressures of discovery claims within a larger scientific community: these are often in tension with the goals of the NASA Mars Exploration Program or its Astrobiology Program as sources of funding for scientific work.

In this chapter, then, I place work with rover images in the broader social and political context of the mission.[4] From this perspective, rover images play an important role in managing the tensions of gaining continual public support for the ongoing mission. This is not unusual in the history of science: images have long been circulated among patronage networks to ensure continued sup- port for scientific projects, whether through sumptuously illustrated astronomy books dedicated to kings and princes, naming newfound celestial objects after prestigious patrons, or circulating other tokens to elicit financial support.[5] Such practices resonate in the Rover team's contemporary context, where the team devotes considerable time and energy to producing special images that circulate beyond the mission to manage relations with NASA and Congress, the planetary science community, or the public. Unlike the maps, annotations, or decorrelation sketches I have discussed so far, these images are frequently spread across the pages of magazines and newspapers, posted on the Internet, and featured in coffee-table books.[6] They fall into a category that art historian Elizabeth Kessler calls "spacescapes": using classical artistic landscape tropes in digital image processing to produce spectacular images for public display that not only capture the public imagination but, in doing so, naturalize political and social relations.[7] Usually heralded with the expression "This is what you would see if you were standing on Mars," the images construct an audience for patronage, appeal to these institutional contexts, and naturalize the rovers' exploration of Mars as a continuing public project. I call this convention *the Martian picturesque*.

I begin with an overview of these three audiences with their concomitant accountabilities and tensions. Then I will describe the conventions and practices of *drawing as* that produce the Martian picturesque, and finally I will situate this image work within the context of the team's external relations.

Audiences and Accountabilities

Although daily work on the Mars Rover mission takes place in a flattened team oriented toward consensus, these social relations are embedded within the funding context of the large bureaucratic agency that is NASA.[8] As described in the introduction, the rovers were funded under the Mars Exploration Program, an office established by presidential decree in the 1990s under NASA's Science Mission Directorate (SMD) that urged a return to Mars to examine the habitability of other worlds. Within the SMD, the Mars Exploration Program competes for funding against the Outer Planets Program, the Lunar Program, and others. Owing to US budget cycles, missions cannot be funded for years at a time: NASA must submit funding requests for these missions to Congress for approval each year. Spacecraft project teams therefore prepare reports on their progress and regularly undergo official reviews to request continued support.

Changes in leadership at NASA headquarters may place missions on the block to make way for new directors with new visions, proposals, timelines, and budgets. When President George W. Bush announced in 2006 that NASA should attempt to send a manned mission to Mars by 2020, this statement boosted funding to centers that focused on manned spaceflight but hurt centers for robotics expertise like JPL, where the Rover operations team members lost many of their colleagues in the resulting layoffs.[9] Even the long life of the mission does not necessarily guarantee continued support. NASA did not expect to be funding the rovers for so many years, nor did it expect the mission to continue alongside subsequent missions such as the Mars Reconnaissance Orbiter (arrived 2006), *Phoenix* (arrived 2008), and *Curiosity* (arrived 2012). This is not only a financial strain; Rover scientists who committed to participating on later missions in sequence have found them running in parallel and must therefore juggle multiple commitments to concurrent projects in their everyday work.[10] With each mission extension there is cause for nail-biting as well as celebration.

Team members' accountability to Congress through NASA makes them constantly aware of the potential funding ax or accusations of inutility that might mean political death for their vehicles. They frequently exhort each other to "give the taxpayers their money's worth on this sol" as they assemble chockablock plans for observations. The PI of the Rover mission is often asked to report to the US Congress on his team's activities and scientific discoveries to justify the continued public expense of operating the rovers, which in June 2008 was approximately $20 million a year.[11] This pressure of continually securing patronage can generate tension when political considerations do not align with local team goals. For example, when characterizing a new region, it is a common practice for geologists to return again and again to areas they have already inspected to build up a more precise regional geological map. However, the opening example in this chapter reveals how this could politically be considered a lack of efficiency. In another case, when *Opportunity* approached Endeavour Crater at breakneck speed and a scientist spoke up to advocate stopping to collect detailed observations along the way to the crater's rim, his colleague countered that with the upcoming NASA-wide funding review and MSL's impending launch, it was more important to reach the crater, with its promise of new vistas and new scientific questions, than to take their time getting there.[12] In such moments, the team frequently identifies scientific and safety rationales that can align with the political impetus so as to avoid cognitive dissonance. As an advocate for the southern winter haven location sighed in frustration, "They're the Mars *Exploration* Rovers, not the Mars *Redundancy* Rovers."

Interinstitutional politics also shape spacecraft operations. NASA is composed of different institutions and subcontractors, each with different relationships with the agency and with each other. Although different NASA centers have different foci, ongoing positioning for agency contracts feeds long-standing interinstitutional quarrels. Thus an ongoing concern on the Rover mission is managing mission members' own institutional boundaries, even while the team attempts to maintain a unified stance and see like a Rover. Images are frequently enrolled in this balance of collectivity and autonomy. For example, visualization experts at one NASA center developed modeling software that could produce three-dimensional environments for rover planning, but another NASA center built its own software and integrated it directly into the rover planning tools. Despite this, certain Rover team members at the first center continue to use the tool in protest. Thus institutional boundaries and commitments remain visible in the visualization technologies themselves.[13]

The Mars Community

Continued funding not only relies on reaching out to NASA and Congress; Rover team members must also demonstrate their continued value to those scientists studying Mars who are not on their team. At the Seventh International Conference on Mars in July 2007, I watched as scientist after scientist from institutions across North America, Europe, and Asia presented their hypotheses about the Red Planet using rover and orbital data, presenting spectral readings, geomorphological interpretations, coregistered overlays, and mineral abundance plots. The scientists NASA selected to participate in the Rover mission are only a small subset of the broader community of planetary scientists who study Mars. They are also just a few voices among the scientists from around the world who regularly participate in NASA's Mars Exploration Program Advisory Group (MEPAG) to advise the agency on which missions to fund that would aid this wider scientific community.[14]

Behind the scenes, then, Rover scientists are also concerned with "generating a really good dataset for the community to mine,"[15] as one Pancam operator described it, and this legacy aspect of rover observations is also deployed in making decisions about which observations to take. When faced with a difficult decision or a choice among observations, Rover scientists often ask each other outright on the line: What would our colleagues expect us to get here? What will people need in the future in order to "do science" in this area?

The team's aggressive image-release policy also reflects this concern for community outreach. Previous missions saw raw and calibrated data carefully guarded by different instrument teams, pored over for science results for publication before being released to the public. In contrast, all rover raw image data is released to the public as soon as it is assembled from downlink, and calibrated datasets are released to the NASA database every three months. This gesture, Rover scientists believe, represents their openness and accountability to their broader community, and it has led to a considerable change in how mission data is made available to the public.[16]

Despite the availability of data, team members must often justify their rationales for data collection to their colleagues. At the Seventh International Conference on Mars, when a presenter called for another sample similar to that acquired at the rock Fuzzy Smith, a team member explained, "We only got one shot at [Fuzzy Smith] and kept looking for another example." When another scientist suggested that the rock called Good Question was named that way because it might be an outlier example in the Independence class of rocks, a team member clarified that the name in fact came from a joke at the SOWG meeting and did not relate to the nature of the material under study.[17] Another outsider confronted a Rover team member after his presentation, claiming "controversy over the interpretation of these structures" owing to insufficient visual evidence: in her experience, the images available online did not provide enough resolution to determine geomorphological properties. The Rover PI replied directly to this scientist's comments by describing *Opportunity*'s high-resolution imaging project at Victoria Crater, explaining that his team was very much engaged in "getting geometries" for geomorphological analysis. And the presenter, a SOWG Chair, also emphasized the situated nature of observation planning:

> It's not trivial that this [imaging] happens; we have to sort of think about it very carefully in the science strategies. For example, when we were encircling Home Plate, we were designing our drive and imaging campaigns very carefully so that we can see each individual exposure of the outcrops as we were going round, and we were not [bypassing] any imaging locations. . . . So yeah, it's not trivial.[18]

Scientists on the mission share a genuine concern for the needs of their broader community, but their deeply embedded and even embodied experience at the time of data acquisition remains crucial to data interpretation.[19] Further, because the rovers are narrative missions that unfold over time, it can be extremely difficult for outsiders to locate observations. As one Rover team member explained it:

> I kind of have heard people [outside the mission] complain. But it's hard. It's not the fault of the people who are on the mission. . . . [There's] some kind of a feeling, you cannot get it from the outside, you need to be in front of a computer looking at data day by day, then when someone mentions a name and you know immediately where that's located and what's around [it], but

for outsiders, they have to look through lots of images. . . . It's harder because you're not inside. You're not in the field.[20]

Addressing this complexity presents tremendous challenges. How does one bring a situated sense of the rovers' activities to those who are not copresent daily?

Public Patronage

Aside from Congress, NASA, and their fellow scientists, Rover team members believe that an audience of amateurs and a generally interested international "public" is following their every move on Mars. As a former Mars Exploration Program Administrator recalls it, the "deliberate decision" to post images online immediately was intended to "build the momentum of public engagement."[21] "People should be able to get up in the morning, get their coffee, log on to the Internet, and see what's happening on Mars today," the Rover Principal Investigator frequently states enthusiastically.

"The public" here is an actor's category referring to an amorphous, invisible, yet extensive imagined audience who may be watching the rovers' progress at any time.[22] Appealing to this audience provides specific benefits, since "public interest" is seen on the mission as a direct appeal to continued congressional funding. But this is a double-edged sword, because nonscientific viewers can easily misinterpret what they see. For example, early in the mission, an amateur on the Internet noticed what looked like a rabbit in a rover photo, and a storm brewed in public forums on the Internet as the meme spread. In 2008 an individual interpretation of a false color image of a rock was trumpeted by international media as the discovery of a Sasquatch or female figure on Mars. The "momentum of public engagement" would therefore have to be balanced by the expertise of public scientists who could interpret the images and provide a scientific *seeing as* experience for this audience of amateurs.

The administrator quoted above described such misunderstandings as "a risk worth taking . . . because it's the scientists who will have the knowledge to really interpret this and who will be the ones up in front of the cameras . . . to stand up and say, 'This is what you're seeing.'"[23] But team members who know how to see like a Rover are often at a loss to explain to a public audience why amateur interpretations are impossible. Members' visual expertise is a tacit, practical knowledge that makes them blind to the aspects of images that naive observers construe as meaningful. Further, it is difficult to exert their visual authority once "public engagement" has achieved any "momentum." Rover members frequently point to other public misinterpretations of Mars images — such as "the face on Mars" or the controversy over evidence of life discovered on the Martian meteorite ALH84001 — to describe the inherent tension between the open attitude of public involvement and potential misunderstandings that may develop and spread.

The possibility of being watched at any given moment and being misinterpreted in their observations inspires a kind of Panopticon mentality:[24] a double consciousness on the team that occasionally surfaces during planning. In one instance, the *Opportunity* team planned an early morning observation of a comet from Meridiani Planum. The observation required waking the rover up early and pointing at a region of the sky during sunrise, not too early but not too late, so as to catch a glimpse of the comet — a tricky observation. After detailing the features of the observation, the SOWG Chair intervened when it came to giving the images a special file name, saying: "I'm not putting 'comet' in the name because what will happen is this will actually end up on the Pancam [web]site and the people who follow along on what we do. . . . I don't want them to look at this and think it's a comet [in case we don't see it]."[25]

It may be true that in case the observation failed (which it did), the Chair did not want the team to look incompetent to their public observers. But the possibility for misinterpretation of the image, for amateurs to think they see a comet in an image in which there is no comet, constituted a greater potential concern.[26]

Among the team's conception of "the public," however, there is a group of amateurs who do regularly watch the rovers: the community on http://www.unmannedspaceflight.com. The approximately 1,500 members of this active online forum discuss the rovers' daily activities, craft homemade algorithms for making their own true color images, and trade thoughts and opinions about the team's rationale for particular actions on Mars. The Rover team is very much aware of this community and interacts with it frequently (fig. 8.1). Team members occasionally check in on the forum's postings to see how they are being interpreted.[27] They also try to guess how the online group will respond to particular activities. Thus the Rover team watches its watchers and maintains a double consciousness about what its amateur public will think of its activities.

This watching reveals an important way that work with images can intervene in some of these tensions between insiders and outsiders, science and exploration, team members and the public. When the online forum's webmaster came to the PI's institution to give a talk to the department about his image work, the

Figure 8.1. Model rover in the laboratory sporting an "I am 2!" birthday pin and card sent by the community at unmannedspaceflight.com. Author's photo.

Rover team members in attendance exuded genuine enthusiasm. After a talk in which the PI exclaimed "Sweet!" and "Cool!" at every image, he finally exclaimed, "I can't tell you how thrilled I am at what you guys are doing. When we made the decision years ago to throw all our images out there it was exactly so you guys could do what you're doing, to follow along . . . do something of substance with them."[28]

The webmaster replied, "Once those images hit the web, I couldn't *not* play with them!" But the "something of substance" that the web participants are praised for doing is very different from the work that Rover team members accomplish with their images. When the webmaster revealed that it took him thirty-six hours to process an image of Mars, the PI pressed, "I know why we do it [work with rover images]; Why do *you* do it?" The webmaster's reply is illuminating: "I guess, take your explanation, why you do it? You do the good science and you do the exploring. We can't do the science but we can do the exploring . . . so we can be right there with you."[29]

Rover Planners might *draw* Martian features *as* hazards for rover interaction, and Rover scientists might *draw* these same features *as* betraying morphological or mineralogical distinctions, but this webmaster appeals to different aims in his *drawing as* activities. And unlike amateurs who might *see* Martian features *as* faces or Sasquatches, the PI characterized this group's work as doing "exactly" what the Rover team had hoped the public would do: to *draw* Mars and *see* Mars *as* a shared site of exploration and experience. Indeed, the webmaster's emphasis on "being right there with you" during the rovers' "exploring" resonates with a team-sanctioned use of rover images, one that deploys this sense of shared exploration in order to manage the complexities of public engagement, NASA patronage, multi-institutional accountabilities, and community service: the Martian picturesque.

The Martian Picturesque

The Martian picturesque begins with a color palette called Approximate True Color (ATC): a choice of filters and algorithms that attempts to imitate as closely as possible the human eye's range of sensitivity to light. Recall that presenting Mars as it would appear to the human eye is not scientifically advantageous. Scientists like Susan and Ben describe Pancam's advantage as due to its multispectral capability and the ability (through *drawing as* practices of selection and composition) to make visible what the human eye cannot see. Another Rover scientist, George, put it this way: "Okay, we know Mars is red, we get it! Seeing more natural Mars colors isn't helping, I'm not learning anything. Seeing Ross's decorrelation stretches? Okay, now I'm learning something new."[30]

But while George claims these images are not useful scientifically,[31] they are produced using considerable time, effort, and resources on Earth and on Mars. Pancam team members in particular work hard to produce Pancam mosaics in ATC for public release. In doing so, they appeal to different techniques and values than they use in their scientific image processing, leaving aside certain constraints but adopting others. Paying attention to their modes of production and their local rationales reveals just what is at stake with this kind of image work.

Composing the Martian picturesque first requires thinking about how to construct a color image that is "true to the human eye." While the Pancam ATC algorithm imitates one model of the human eye, other true color algorithms can vary. George, for example, challenged the Rover team's convention on aesthetic grounds rooted in his sense of public interest:

> [My Pancam colleague] always wants to make Mars images look dark because if you were there you're farther from the sun so it'd be dark. . . . It's the gloomiest, saddest, most depressing [view] you could imagine. . . . If you were on Mars, your eyes adapt. Yes, Mars is darker, but my eyes would adapt because my pupils would dilate. So why are you making it gloomy when it doesn't have to be? . . . Even with visible imagery there's real differences of do you make it look exactly like you were standing there or make it look something like, okay I'm attracted to this image, I want to look at it, I want to peer into the shadows.[32]

In George's view, the appeal to the human eye not only is a question of anatomical accuracy, but also is about public engagement and attraction. George is particularly thoughtful about the aesthetics of data display, a sensitivity he credits to studying the work of data visualization expert Edward Tufte. George's research team hires an artist to help his group produce images that captivate the public. According to George, material released to the public has "gotta be pretty, but it's also gotta be intuitive." In addition to his participation on the Rover mission, George also works with an orbital instrument that produces nonvisual data, so the team could adopt any palette to more readily depict its results. He told me about his discomfort when he found that public enthusiasm was much greater when their images were released with a Marslike palette: browns, oranges, reds, and butterscotch. In our interview, George presented this as an ethical dilemma: "Your eyes can't even see these wavelengths. . . . Should we be putting Marslike colors on something that's infrared data?" As the appeal to "what it would look like if you were standing on Mars" migrates even to the depiction of invisible data, this reveals the importance of color palette in an appeal to the public (fig. 8.2).[33]

Mars Rover Pancam team members spend considerable time and effort making Pancam image mosaics in ATC for public release, making sure their images produce a standardized view of Mars. Early in the mission, Pancam mosaics were

Figure 8.2. Thermal data of Mars displayed in a butterscotch color scheme, like visible light. Chasma Boreale (Christensen et al., "THEMIS Public Data Releases"). Courtesy of NASA/JPL/ Caltech/Arizona State University.

stitched together one frame at a time through image-processing software, and true color images were produced by hand coding the appropriate transforma- tions and making adjustments by hand in software. Since then these processes have mostly been automated to ensure a more unified view of Mars across the board,[34] but sometimes the algorithm results in an image that doesn't look quite right and must still be adjusted by hand. In one case I witnessed a Pancam mo- saic maker shake his head at an image that came through the software pipeline, saying "It's too red!" and opening Photoshop to adjust individual properties until Mars was less "red."

Another challenge arises from the Pancam itself. By the time the camera physically rotates from the left side of a panorama to the right, several hours, if not days, may have passed between individual frames. In the meantime the brightness and contrast of the Martian sky and terrain may have changed from frame to frame, so that different panels of the mosaic can present different col-

Figure 8.3. The Martian picturesque featuring Approximate True Color (ATC) with just the right balance of red and equalized sky color, with tracks receding across a gentle landscape near the Home Plate region. *Spirit* sol 613. Courtesy of NASA/JPL/Cornell.

ors for ground and sky. To minimize the discrepancy, the team member who is crafting the mosaic may adjust the frames so the ground is a consistent color, then select the color of the sky at one point in the image and paint over the rest of the sky in the scene with that color, creating a uniform sky and ground. Note that while painting over a scene would be considered "lookiloo" in a scientific context, this is not the case for public release images. Because the sky value is an actual pixel value from the sky, as one mosaic maker explained while making the change, "you're not inventing values" (fig. 8.3).

In addition to color considerations, images that deploy the Martian picturesque pay considerable attention to framing. When planning Pancam mosaics, Pancam operators—many of them accomplished amateur photographers and artists themselves—may speak up to represent aesthetic considerations. For example, Thomas explained that if he thought the imaging sequence needed "one more frame to make it prettier," he would simply ask on the SOWG line if that was okay "within the limits of our resources," and it was generally approved. "Making it prettier" involves thinking about how the individual Pancam images will stitch together to create a larger picture. If two scientists suggested two observations of two separate objects that were close to the surface and involved the same exposure, Thomas might suggest an additional frame to "make a nicer

Figure 8.4. The Martian picturesque, again featuring tracks snaking off with the spectacular vista of Victoria Crater in the distance. Note how the cluster of meteorite debris nearby forms part of the mosaic's compositional element by balancing the swerve of the tracks. These Pancam images were taken as part of a test for an upgrade of the navigational software. *Opportunity* sol 1162. Courtesy of NASA/JPL/Cornell.

picture because it's all together." He might also suggest a particular time of day or combination of filters to capture qualities of the light and "improve" on the requested image. "Generally everyone wants the prettier image but within constraints," he offered. "I do it to make it look nice, and generally the scientists care about that too."[35]

A primary characteristic in framing reflects a genre broadly classifiable as that of the American frontier. Countless rover images show tracks receding toward a distant horizon reminiscent of wagon wheels on a pioneer trail (fig. 8.4). The award-winning promotional animation for the Rover mission (produced by a student on the mission at the time) depicts the rover descending from its landing module onto the Martian terrain and heading off into the sunset like a cowboy in a Western movie.[36] Such framings frequently evoke American photographic traditions ranging from Ansel Adams's towering landscapes of Yellowstone or the Grand Canyon to nostalgic views of the western frontier. These aspects appeal to the "exploration" side of the mission, which may conflict with the slow and steady work of science and rover management described above, but they also present a familiarly American view of the Martian landscape, transforming Mars into the new frontier.

The frontier narrative is visible even in a cursory viewing of public Mars rover images. But the genre is also at work behind the scenes. The Pancam panoramas at Victoria Crater described in chapter 1 are excellent examples. At first approach to Victoria, one scientist said he was overwhelmed by "the sheer absolute phenomenal beauty of the scene," and he requested a panorama, saying, "That's obviously not a scientific driver but something that's always in the back of our minds."[37] Recall also the SOWG Chair's request for a panorama of Victoria Crater "with the sun low in the sky," the shadows of the promontories reminiscent of the Grand Canyon.[38] When advocating for the observation, as I described in chapter 1, the Chair drew an association between an aesthetic and the audience: although "some good" (something useful) might come out of it, neither science nor operations is the image's primary function. He invokes traditions in American western landscape photography, citing "postcards" and familiar images of the Grand Canyon to claim he is doing the same sort of thing. After the meeting a team member laughed that the Chair was "playing Ansel Adams,"[39] and the panorama was known as the Ansel Adams Pan from then on.

The proposed image was meant to be "spectacular," "a postcard"—or as the Chair put it later, "It's not science, but it'll be cool!" In chapter 1 I described how this talk in a SOWG meeting reminds team members of their collective exploration of Mars. But the audience for this image is not the team. Instead, it "could become the image of the week" displayed for public engagement on a NASA website. In this moment the public and the aesthetic are rhetorically intertwined and realized through the planning and execution of the photograph.

The Martian picturesque is participatory. Transforming Mars into a vision you would see if you were there invites the viewer to step into the frame, into the rover's tracks so often visible in the scene. This is not a view from nowhere or a God's-eye view. Instead the viewer is very clearly situated on Mars, alongside the robot. Nor is it especially a rover's-eye view. Unlike the conventions of seeing like a Rover described in chapter 6, these images present a view oriented toward the human observer. This was especially underscored for me when I interviewed a Pancam team member who was attempting to write software that would convert all rover images to a perspective as if the image had been taken from six feet above the Martian soil, not five. This, he explained, would be more like a human's perspective on Mars than the rover's.

Locating viewers in a stark landscape with a scene laid out around them also recalls the picturesque convention in eighteenth-century landscape painting. Usually associated with the pastoral landscape, the picturesque elevates the everyday into scenes that are calm, peaceful, and composed. It arranges the landscape around an observer who is embedded within it at a particular location. The viewer is not overwhelmed by their surroundings such that the view

Figure 8.5. The Martian picturesque, featuring a landscape slightly askew, rover panels and mast visible in the scene, single sky color, and rover tracks. Compare with figures 5.6 to 5.8 of the same region, used in the scientific and engineering discussions at this time. West Valley Pan, *Spirit* sol 1366. Courtesy of NASA/JPL/Cornell.

is terrible, awesome, or emotional (as in the sublime); nor are they observing from an impossible vantage point. Instead, they are grounded and embedded in a scene that is peaceful and tangible, domesticated yet enchanting, occurring at a precious place and time.[40] One must be in the right spot on the ground to enjoy the picturesque view, to have the elements of the countryside arrange themselves just so.

Such conventions resonate in these twenty-first-century examples. Color and composition are combined to present the rover's-eye experience of the alien world on a human level. The situated nature of the robotic viewer in Martian picturesque imagery is often highlighted by foregrounding a panorama with splayed rover solar panels or framing tracks visible in the sand, reinforcing the position of the subject observer as rooted in the scene and producing the sense of the landscape as slightly but charmingly askew, a "found" moment in an untouched space (fig. 8.5). Human presence on Mars thus appears natural and seamless, arising from the landscape. This is the result of members' practical image craft that draws Mars as the new American frontier. It is the visual aspect of the Martian picturesque.

The predominant aesthetic considerations at play in producing the Martian picturesque reveal that these are not scientific images. There is no concern here about replicability or mathematical expressions, constraints based on experience on Earth, or making new features "pop out." They may serve a purpose as a marker of a significant team achievement, as described in chapter 1, but I never saw these images analyzed for scientific features. Thus the significant time, effort, and resources that go into producing them require an alternative explanation. In my experience, these images were rarely produced without some invocation of "the public."

Drawing the Public Together

In earlier chapters I have shown how *drawing as* work with images not only configures visions of the planet, but also manages social relations within the team. Here I address how the team members seek to deploy the Martian picturesque to manage their external relationships. The institutional environment of the Rover mission reveals a range of institutional pressures, expectations, and concerns. Managing this patronage network is a complex and even contradictory venture. What political and social relations are naturalized in this symbolic landscape? How do these representations of Mars—Mars *drawn as* the frontier, as if seen by the human eye—ease the various tensions described above?

The most obvious function of producing the Martian picturesque is for public relations documents: images that remind "the public"—and through this public, Congress—of the continued value of their mission. Far from denigrating such images as a "dog and pony show" distinct from scientific study,[41] Rover team members are passionate about these images and rhetorically associate them directly with mission success. The Pancam Lead once declared to a classroom of students, "It would be a crime against humanity to send a spacecraft without a camera,"[42] pointing not to the scientific merits but to the public imperative. George matter-of-factly stated, "How many people know we have two rovers on Mars? I'll bet you nine out of ten people know that, and it's because of the images."[43]

Images therefore stand in as a measurement of mission success: they are the most easily perceived and shared "deliverable" of the mission. This was perhaps best articulated by members of the earlier Mars Pathfinder science team, many of whom are currently Rover mission members, when their camera was threatened by budget cuts in 1994. In a protest letter, they said,

> Try to imagine two successful Viking landings on Mars in 1976 followed by no images, no samples, and no sample analysis to test the hypothesis for life on Mars. Try to imagine the successful landing of *Apollo 11* on the Moon with only voice communication—no pictures, no samples, and no televised "first step." It is important to recognize that images from the surface of Mars will prove success to the American public (and Congress) and provide them with tangible results they can comprehend.[44]

Figure 8.6. Panorama in ATC acquired at Duck Bay. Note the Photoshopped rover placed in the picture "for scale." Courtesy of NASA/JPL/Cornell.

Pictures of Mars taken by a robot are here equated with the world-famous televised landing on the moon and Viking lander pictures in terms of their scientific and emotional impact. The pictures themselves are said to "prove success" and provide "tangible results" that taxpayers can "comprehend" by simply seeing them. This does not imply that taxpayers will see the pictures and judge for themselves the geology of the scene around them; rather, sharing the imagery is itself a "tangible result." It also produces "tangible results" for the team, since public appeals are equated with more possibilities for public funding.

The use of frontier imagery especially appeals to the American Congress, public, and NASA alike with a shared cultural understanding of these images as standing for the greatness of their nation and their accomplishment. Such visions of Mars domesticate the planet and naturalize continued support and patronage of the mission. But at the same time, they reach across many audiences with divergent concerns with their appeal to "what you would see if you were standing on Mars." The most important word in that sentence fragment is the pronoun "you," which notably replaces the "we" that team members most

frequently use behind the scenes. This produces a strikingly different rhetorical appeal with a different network of social relations.

In this respect the stance of the observer within the Martian picturesque is critical. This became clear to me when the Pancam Lead presented the panorama of Duck Bay at a NASA press conference in September 2006. In addition to the panorama—not the version we had all seen behind the scenes earlier in the week, but one cleaned up and specially processed to ensure accurate colors and resolution—he displayed a version that placed a Photoshopped rover in the scene: as he described it, "for scale" (fig. 8.6). Looking at the image of Duck Bay with a rover placed atop Cape Verde, however, was disorienting. As a socialized member of the team, I was used to seeing like a Rover, looking through the robot's eyes at the terrain with all the embodied and social work that implies. With a rover imposed on the scene, suddenly I was standing on Mars alone, outside the "we" of the robotic body, looking at the rover looking at the crater.[45]

The cognitive dissonance I felt at this moment was the experience of a change of aspect owing to the imposition of a new visual convention: a new *drawing as* practice that produced a new *seeing as* practice for the observer. Instead of turning human bodies into robotic bodies, the Martian picturesque transforms the rover's visual production into a human visual experience. The stance of the observer in the rover's tracks at America's new frontier, the transformation of Martian imagery into a color palette that appeals to the human eye, and even elements "for scale" combine to craft a particular kind of virtual witnessing experience for the viewer. A crafted and intentional image feels like an individual observation, and the alien planet is rendered familiar and knowable to distant human observers, forging a connection with Mars. The edges of solar panels visible in the frame, or the occasional "self-portraits" of the rovers themselves, remind viewers exactly which robot and which mission is making this vicarious experience possible.

At the same time, a more subtle appeal in the Martian picturesque is the ability to reach across the conflicting and contradictory boundaries and networks of accountability that engulf the mission. Certainly, such images provide a sense of place for other Mars scientists attempting to use rover data in their own work. They thus bring the outsiders in and place them into the scene. But the use of the frontier framing and genre in the images' narrative also attempts to bridge divides outside planetary science. Ross explained this connection when discussing what he felt was the overall importance and impact of the mission: "Doing planetary and space exploration I think really helps society, giving us a frontier, a place to push our boundaries. . . . The problem with American society is we don't have a frontier anymore, so we're turning on each other."[46]

Although frontiers frequently stand as places of confrontation between cultures, indigenous or otherwise, or as spaces of violent conflict, Ross evokes the Martian frontier as a way to overcome divisions. His comments were framed with a discussion of the difficulties in American society as he saw them: a polarizing of issues and divergence between left and right ranges of the political spectrum. Frontiers require rugged and hardy explorers who exert creativity and ingenuity to manage their environs, but they also require collective political will. A shared sense of "pushing the boundaries" enables a more harmonious "American society" that can agree on at least one thing: the importance of exploration to national pride and predominance.

Their scientific virtues aside, then, true color images play a significant role in the continued success of the Mars Rover mission through their positioning as circulating objects that entice and enlist external viewers into support of the mission. Through the use of Approximate True Color, frontier resonances, and a view from behind the rovers' eyes, such images *draw* Mars *as* a new frontier, at the same time as they put viewers within arm's reach of this other planet.

The standpoint of the virtual witness is thus a political stance: it unites multiple bodies through a robot on Mars and requires their renewed commitment to this robot. These images do not just invite viewers to imagine themselves standing on Mars, to perhaps become astronauts someday. With an appeal to "what your eye would see," the Martian picturesque presents "postcards from Mars" that aim to draw together a variety of publics to which the rovers are accountable. As such, the Martian picturesque bridges the individual tensions inherent in each organization and group and presents instead a unifying vision with a direct appeal to the observer—the "you" in the frame.

Epilogue: Surviving Politically

It is beyond the scope of this book to show how public images are received or whether this *drawing as* strategy is successful in invoking the *seeing as* vision with its full social implications. But a moment during my fieldwork suggests that these images establish strong relationships between various public stakeholders and the rovers themselves and can mediate between or sidestep contradictory accountabilities.

In March 2008, NASA's Associate Administrator for the Science Mission Directorate's office issued a letter to the Mars Exploration Program denying requests for increased funding. The new rover, *Curiosity*, was officially over budget, and the Directorate decreed that the Mars Exploration Program would have to manage the cost overruns on its own. The program could not support the needs of the Mars Science Laboratory alongside the two rovers and the Mars Reconnaissance Orbiter with the amount of money in hand. The Mars Program Administrator delivered a letter to the Rover team indicating that it had to cut $4 million from the current fiscal year's operating budget, and up to $8 million for the next year; a similar letter gutted MRO's resources. Reviewing the proposed 20 percent cut to his team, the Rover PI responded that there was no way to operate two rovers for that amount: they would have to shut *Spirit* down. The mood in the lab was glum as the changes were announced on the SOWG line, then broadcast through press release on Monday, March 24, 2008. It seemed the rover had survived physically, but not politically.

The press release hit news websites on Monday at noon. By that afternoon, websites from cnn.com to space.com to physicstoday.org teemed with outraged comments from "the public." The webmaster of unmannedspaceflight.com e-mailed the Rover team to reiterate his community's continued support for the mission and offer any help through letter writing or other activist ac-

tivities that could reverse the funding decision. CBS ran a story detailing "THE
OUTRAGE FROM SCIENTISTS TO SWITCH OFF SPIRIT AND RUN OPPORTUNITY
EVERY OTHER DAY,"[47] while a commentator on spacepolitics.com simply stated,
"I would venture to guess more people could name both rovers on Mars than
could name a single member of the current astronaut corps."[48] The website i09
.com, frequented by science fiction fans, posted an article about *Spirit*'s "death
sentence by the U.S. government," claiming, "To say that this is a tragedy is an
understatement." Next to the blog post, above the comments, was a true color
image of the Martian landscape, rover tracks winding off into the horizon, cap-
tioned as "a picture *Spirit* took of its own tracks in the dust."[49]

The story moved too quickly for the mass media. As members of the team
who lived on the East Coast made their way home on Monday afternoon, the
question whether the Jet Propulsion Laboratory would stand idly by to watch
its prized mission canned by a Headquarters decision had already been raised.
By Tuesday morning, NASA administrator Michael Griffin announced that the
budget letter would be rescinded: NASA would not kill one of the rovers. On
Wednesday morning NASA announced the Associate Administrator's resigna-
tion. "Did the Internet just accomplish something?"[50] mused one blog com-
menter when the story made its way to universetoday.com.

If "the Internet accomplished something" that afternoon, it was as the space
for "the public's" overwhelming response to the rovers' political plight. It was
also the space where over four years of images had been publicly released straight
from the rovers' cameras to individual desktops across the country. As a result,
"the public" that the Rover team so frequently invoked had come to see the
robots as its own, had developed a relationship with them and their journey,
and experienced a sense of copresence at the new frontier. Four years of *drawing*
Mars *as* a human experience at the new American frontier was likely in no small
way responsible for the public reaction, and as a result the administrator's quick
action to rescind the letter. What "the Internet" also "accomplished," however,
was to act as the site for an institutional contestation and an escalation of concern within another organization to which the rovers were accountable: NASA.
Unable to cut mission spending but still facing the mandate to satisfy budget
restraints in the face of MSL overruns, the Associate Administrator resigned.[51]
The rovers, at least, had deflected the threat through an appeal that mobilized
another of their patronage groups, united around the sense of personal connection and experience that the Martian picturesque evokes.

"We're planning with a lighter heart than I expected," the Chair exclaimed at
the SOWG meeting that Wednesday morning. *Spirit* had survived yet another

near-death experience, this one political rather than physical. But before getting down to the business of the day, the PI piped up on the line with a reminder of the robot's precarious position on both planets, and therefore also the team's continued attention to its local social order. He sounded both relieved and cautious: "The only thing I would add is, we don't know what's gonna happen next, so live for the moment. Get all you can out of this rover today, guys."[52]

CONCLUSION

Spirit's third winter on Mars was her last. After a spring and summer season of observations on the north side of Home Plate, the robot drove across Home Plate toward a fourth winter haven site to the south. En route, a wheel broke through a delicate crust layered on top of the ancient hot spring. Spirit was trapped. With each drive command, the rover dug herself deeper into the soil. When the team commanded the robot to use her IDD arm and Microscopic Imager to take a picture of her underside, the fuzzy image that returned showed that she had also lodged herself on top of a rock, impeding her ability to drive away.

On Earth, the team worked frantically to free its robot, or at least tilt her so that her solar panels would soak up enough sun to survive the winter. Mark, Sarah, and other Rover Planners spent endless days in the test bed at JPL, working with their model robot to simulate the conditions on Mars. Nick and Katie each flew out to JPL in turn to lend a scientist's eye to the proceedings, picking out the appropriate soils with the right qualities to best approximate *Spirit*'s location.[1] JPL ran a web campaign called "Free *Spirit*," and Rover staff sported the logo on T-shirts and buttons. Promising results came through at the end of January 2010, but by then NASA headquarters had already decided to stop funding dual rover operations. The mood went from frantic to excited, then transitioned to anger, anguish, and grief.[2]

March 22, 2010, was the last time the team heard from *Spirit*. The mission managed to secure resources to continue listening for the rover throughout the winter and into the spring, in case of what they called a "Lazarus situation:"[3] a miraculous revival once solar power went up in the spring. Based on their power models, the team hoped to hear from her again in April or May at the latest. No signal was received, however. On May 31, 2011, NASA Headquarters declared the recovery effort over. *Spirit* was dead.

In July 2011 the Rover team met in Pasadena for the usual Team Meeting. *Opportunity* was only a few hundred meters from its newest goal, Endeavour Crater, and would be there within the week. Tom's DEM maps of the crater based on HiRISE and spectral signatures from orbit were projected on-screen alongside Joseph's geological sketch maps, Peter's drive plan sketches, and Ben and Ross's colorful Pancam images. But the team was also there to celebrate their lost rover, now memorialized with the naming of *Opportunity*'s arrival location at Endeavour Crater: Spirit Point.

That evening, over a barbeque at the Caltech Athenaeum, about sixty scientists and engineers gathered on the lawn: the engineers who had built the rover alongside those who had steered her across Gusev Crater, and the scientists who had used her instruments. The PI's comments were brief, focusing on *Spirit*'s longer-than-expected life and her triumphs on Mars and crediting above all the extraordinary team that built her, drove her, and conducted science with her every day on Mars. In death the rover had once more become inanimate metal, the source of whose animation was clearly the people gathered around the picnic tables.

Touring the tables, I asked team members for their favorite stories about *Spirit*. Sam recalled his naming Innocent Bystander when *Spirit* crushed the wrong cobble. Susan cheered the discovery of Tyrone and the "amazing coincidence" of spatial and spectral information using Pancam. Nick, relating his euphoric moment when he realized there might be a relationship between Gertrude Weise and biologically formed sinters on Earth, compared the high of discovery to "being on drugs" and then described his eventual realization that it was not the case. Adam recalled a moment in an airport when he realized that at that very second a robot was doing what he told it to do on Mars, moving its wheel "just so." As he spoke, he twisted his wrist and arm, bringing the rover's movement into his own body. As these team members recalled their memories of *Spirit*, they spoke of moments not only on Mars, but on Earth too.

This book has sought to make a similar interplanetary connection. The visual work of taking and analyzing pictures of Mars is also a question of work on Earth, managed through interactional moments, rituals, and social norms. Each image and each transformation of an image produces a kind of scientific seeing, revealing particular insights about the Red Planet. At the same time, I have argued, interactional norms around image taking and image processing reinforce and reproduce local conditions of work. The work of managing the rovers is also the work of managing the team.

These two activities are inseparably intertwined. They are two sides of the same coin: the duck and the rabbit in the gestalt image. Notably, this is not an either/or relation, but a both/and. After all, Rover team members would never say "Now we're doing knowledge work, now we're doing social relations." Both aspects are present at the same time in the work of Mars Rover mission team members as they work with the digital visual materials that return from Mars. Within the framework of *drawing as*, the duck/rabbit does not represent analytical ambiguity so much as analytical opportunity.

If such aspects are both present, however, they are revealed by my own work of *drawing as*. Throughout this book I have used a few key examples to show several aspects of the team's local order. In the cases of Winter Haven 3, the many transformations of Tyrone, and the circumnavigation of Victoria Crater, I hoped to show how many different organizations of my own fieldwork experience are possible even in a relatively bounded study of a scientific laboratory at work. Lest this seem like a plea for analytical relativism, however, note the close attention throughout to issues of skilled practice. Drawing a natural object as an analytical object requires work to assemble elements of the field and organize them so they "pop out." It requires bodies and body work (chapter 6), interactions according to local norms and rituals (chapters 1 and 5), attention to institutional arrangements and organizations (chapter 8), the display of disciplinary perspective (chapter 4), and adherence to disciplinary norms (chapters 2 and 7). Maintaining our focus on practice reveals the importance of skilled work in producing these particular accounts and aspects, whether in planetary science or in the sociology, history, and philosophy of science and technology.

I opened this book with an appeal to questions of representation in scientific practice. Although we often believe that representations stand between an observer and the world, this study demonstrates how images of the world also represent an observer's work in the world. Working with images of Mars involves seeing, drawing, and interacting as iterative activities, each inspiring and contingent on the others in the unfolding narrative of robotic exploration.

In their myriad interpretations and projections, Rover team members employ images as a resource to both conduct their science and manage their community. Conducted with materials ready to hand and with robots millions of miles away, this work is at the same time practical, technical, social, and epistemological as it makes Mars available for interaction. This team and its interactional and organizational work with image making and interpretation is particularly oriented toward the continual production of consensus, hierarchical flattening, and a concomitant social order. The wide variety of interactions that render Mars workable and meaningful to its Earthbound observers are embedded in the many images produced by the Mars Exploration Rover mission: from the raw frames freshly downlinked to the Internet, to the Warholesque false color prints in scientific journals, to the fold-out panoramas in coffee-table books.

In the analytical phrase *drawing as*, then, I propose a synthesis — or perhaps a drawing together — of current formulations about representation in scientific practice into a suggestive way to formulate image work: as simultaneously the site and document of knowledge production in the sciences. Even as scientists employ and invent visual languages for categorizing objects of interest,[4] they exert their professional vision[5] over the image and inscribe their discrimination of categories,[6] sense-making practices,[7] and meaning into the image itself. To do so credibly requires an appeal to a local formulation of objectivity[8] such that the image can be *drawn as* trustworthy and *seen as* "evidence of anything."[9] This process requires exerting discipline both over the pixels in the image and over the pixel-pushing scientists themselves,[10] conscripting eyes, hands, and machines in careful coordination to produce trusted images and communities of scientists alike. With each twist of the storyline, however, the action may be further analytically from the observation's original evidentiary context.[11] The analytical phrase *drawing as* therefore takes an empirical and material approach to describe how interactional practices with graphic materials inscribe traces of an object's analytical production into the image, documenting scientists' work "in action" with the visual data they interpret. It therefore points not only toward a practical activity available for accounting, but also toward an activity that leaves graphic traces even as it shapes how objects are appreciated, interacted with, and seen.[12]

This raises the question of the applicability of the analytical frame to other studies of scientific visualization. Although my study focuses on a site of digital image work, I emphasize thematic continuities between practices of representation across different historical periods and media. Like Mars Exploration Rover images, Galileo's cratered moon and Lowell's Mars crisscrossed with canals, mentioned in chapter 3, are also examples of *drawing as*: inscribing categories and distinctions into the image of the object and embedding a way of *seeing as*—a way to appreciate that aspect—into its very representation. Studies of images in the history of science such as representations of nebulae, cosmological systems, and various other astronomical phenomena similarly describe how theoretical claims and analytical distinctions about kinds of objects and their meanings are drawn into images of the day.[13] Nor is *drawing as* unique to astronomy. Eighteenth- and nineteenth-century anatomical illustrations can be analyzed for traces of historical orientations toward gender and sexuality;[14] geologists in the nineteenth century drew debates about catastrophic change into their paleontological images;[15] and even Feynman diagrams express communities and changes in theoretical physics.[16] In these and other cases, modes of seeing are impressed onto the canvas and are taken up with the image and with the object as matters of fact.

In addition to this historical orientation, *drawing as* opens up questions for analysts of contemporary work with digital images. Although much of "where the action is" has moved to the screen, to computational algorithms, or to teleconference lines, in this virtual space images require work to make them present, accountable, and traceable. The "externalized retina"[17] does not disappear from the laboratory but continues to be highly situated, implying shared modes of viewing and confrontation with an alien frontier. The practices that constitute this digital visual work have been particularly well documented in studies of false color, image composition, and gesture while working with brain imaging technologies,[18] although their applicability to the practice of planetary science has received limited attention.[19] The ability to combine and recombine the same image in so many different ways, and to bring these versions into conversation with one another, presents an opportunity to witness *drawing as* in various places and stages of action. This may afford access to multiple partial perspectives[20] and possibilities for exhibiting and addressing notions of incommensurability.[21]

If *drawing as* constructs knowledge of Mars for the Rover mission members, however, it does so only insofar as it is embedded in the social order adopted and reinforced among team members, which is essential to knowledge making: the form of life in which science is conducted, observations are produced, and hypotheses gain validity.[22] This book has argued that work with rover images draws the scientists of the Rover team together as members of a social and micropolitical body, exhibiting expressions of solidarity around shared values of collectivity. This solidarity is mediated, expressed, and exhibited in the production of their images, which at the same time provide a focus for the team's communal seeing as practices that tie members to the rover and to each other. Understood in this way, the practices of *drawing as* produce the intersubjective activities of seeing like a Rover, supporting the team's interaction rituals and political structure. Recalling Liz's assertion that "after those rovers leave Earth, the team is all we've got," it is at least clear that images of Mars produced by the Mars Exploration Rovers reveal as much about the Rover team as they do about the Red Planet.

Images of objects are images of subjects too. Therefore the analytical questions we ask of image work must take into account the kinds of social relations in which representers and audiences are embedded.[23] After all, knowledge making occurs in communities. Such communities may present examples of disciplinary visions,[24] of distinctions based on technological skill or division of labor,[25] or of bureaucratic visual production that imposes a social ordering on the world.[26] In the Rover team, we have an example of no less total ordering, but in this case we see a flattened hierarchy, a celebration of disciplinary diversity at the same time as there is struggling with flattened distinctions on variables like seniority and partial perspective. The struggle to see like a Rover, as if it could ever be a unified point of view, is a case in point. Crafting this as a singular view takes considerable social work.

But this social work is also largely achieved through image work. The interactions of working with images are the interaction rituals of the team that keep the group together, enforcing and reproducing their particular social order. Images in interaction, in all their various interactional modes, produce both social (and robotic) relations and, at the same time, the epistemic distinctions that make Mars visible and knowable. This points to a central role for images in scientific practice. After all, it is fluency in all these interactions that brings team members together, in the same situated position on Mars, to collectively see like a Rover.

APPENDIX A
Traverse Maps

Spirit

Spirit landed in Gusev Crater on January 3, 2004, at the Columbia Memorial Station at the upper left corner of the image (fig. A.1). After briefly visiting Bonneville Crater, the rover drove across the crater floor to climb the Columbia Hills, on the right side of the image, where it spent its first winter on Mars. From there *Spirit* proceeded south to the area called Home Plate, pictured in detail in a HiRISE image (fig. A.2).

Fieldwork for this project took place during *Spirit*'s exploration of Home Plate. In the image from the orbital HiRISE camera one can see the distinctive shape that gives Home Plate its name. To the right is Mitcheltree Ridge; the area of activity west of it is Silica Valley, the location of the silica-rich Innocent Bystander and Gertrude Weise. The extent of *Spirit*'s tracks to the bottom right of the image is Tyrone: under the yellow traverse lines the white soil can be seen from orbit. South Promontory is to the bottom left, and the disputed winter haven ("WH3") is at top left.

Figure A.1. *Spirit* regional map: Gusev Crater. Image released December 3, 2007. Image credit: Ohio State University Mapping and GIS Laboratory/NASA/JPL/Cornell/Malin Space Science Systems.

Opportunity

Opportunity landed January 24, 2004, in Eagle Crater on Meridiani Planum and has since explored a variety of craters in the area, proceeding from one to another (fig. A.3). Fieldwork for this project took place as the rover explored Victoria Crater, beginning at final approach to the crater and concluding as the rover moved south toward Endeavour Crater.

   Opportunity arrived at Victoria Crater in September 2006, at Duck Bay, and proceeded clockwise around the crater to the dust streaks on the upper right of the image (fig. A.4). This close-up of the rim of the crater taken by the HiRISE orbital camera is annotated with the names of the promontories and the rover's tracks up to sol 1188 (end of May 2007). Based on imaging conducted at the

Figure A.2. *Spirit* local map: Home Plate. Image released November 16, 2007. Image credit: NASA/ JPL/Cornell/MRO-HiRISE/NM Museum of Natural History and Science Systems.

promontories, *Opportunity* returned to Duck Bay and entered the crater there. Cercedilla is on the edge of Golfo San Matias near the Cape of Good Hope, upper right. The rover is visible in the orbital image on the promontory at Cape Verde (lower left).

All traverse maps in this appendix are publicly released and available online at http://marsrovers.jpl.nasa/gov/mission/traverse_maps.html.

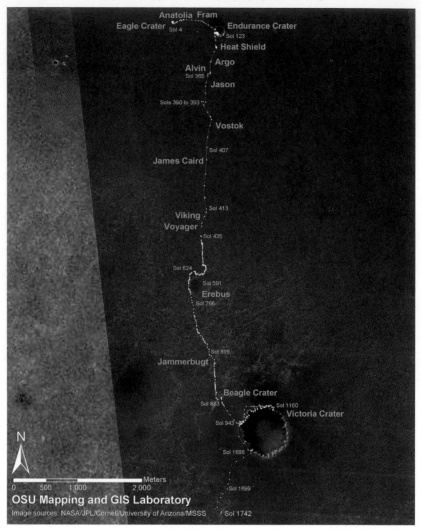

Figure A.3. *Opportunity* regional map: Meridiani Planum. Image released December 29, 2008. Image credit: Ohio State University Mapping and GIS Laboratory/NASA/JPL/Cornell/University of Arizona/Malin Space Science Systems.

Opportunity Traverse Map (Sol 1188)

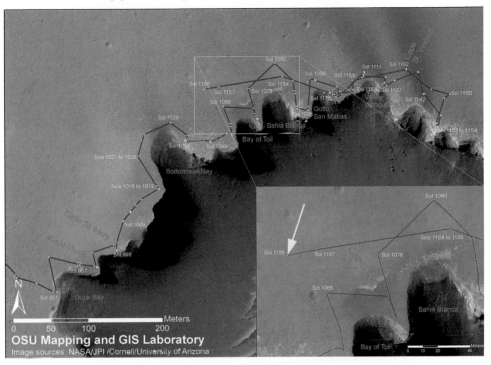

Figure A.4. *Opportunity* local map: Victoria Crater. Image released May 30, 2007. Image credit: Ohio State University Mapping and GIS Laboratory/NASA/JPL/Cornell/University of Arizona.

APPENDIX B
Fieldwork

Participants

All participants appear in this text under pseudonyms. However, I am extremely grateful for interviews, observations, lab tours, and other extended conversations with more than eighty participants in my study, from undergraduate students to senior personnel, including the following:

Oded Aharonson, California Institute of Technology
Ray Arvidson, Washington University, St. Louis
James Ashley, Arizona State University
Don Banfield, Cornell University
Charlie Barnhart, University of California Santa Cruz
Shianne Beers, Cornell University
Jim Bell, Cornell University
Ross Boyer, NASA Ames Research Center
Diane Bollen, Cornell University
Natalie Cabrol, NASA Ames Research Center
Wei Chen, Ohio State University

Phil Christensen, Arizona State University

Barbara Cohen, NASA Marshall Spaceflight Center

Larry Crumpler, New Mexico Museum of Natural History and Science

Emily Dean, Cornell University

David Desmarais, NASA Ames Research Center

Kaiching Di, Ohio State University

Doug Ellison, www.unmannedspaceflight.com

Bill Farrand, Space Science Institute

Paul Geissler, US Geological Survey

Amitabh Ghosh, Cornell University

Trevor Graff, Arizona State University

John Grant, Smithsonian Institution

Ron Greeley, University of Arizona

John Grotzinger, California Technical Institute

Ed Guinness, Washington University, St. Louis

Shaojun He, Ohio State University

Ken Herkenhoff, US Geological Survey

Scott Hubbard, SETI Institute

Ju Won Hwangbo, Ohio State University

Byron Jones, Jet Propulsion Laboratory

Jeff Johnson, US Geological Survey

Jonathan Joseph, Cornell University

Laszlo Keszthelyi, US Geological Survey

Kjartan Kinch, Cornell University

Thomas Kneissel, Freie Universität Berlin

Amy Knudson, Washington University, St. Louis

Alistair Kusak, Honeybee Robotics

Geoff Landis, NASA John Glenn Research Center

Kevin Lewis, California Institute of Technology

Ron Li, Ohio State University

Kim Lichtenberg, Washington University, St. Louis

Justin Maki, Jet Propulsion Laboratory

Scott Maxwell, Jet Propulsion Laboratory

Elaina McCartney, Cornell University

Timothy McConnaughy, Cornell University

Tim McCoy, Jet Propulsion Laboratory

Patrick McGuire, Washington University, St. Louis

Chase Million, Cornell University

Jeff Moore, NASA Ames Research Center
Mary Mulvanerton, Cornell University
Gerhard Neukum, Freie Universität Berlin
Eldar Noe, Cornell University
Jeff Norris, Jet Propulsion Laboratory
Cindy Ota, Jet Propulsion Laboratory
Oleg Parisen, Jet Propulsion Laboratory
Gale Paulsen, Honeybee Robotics
Sylvan Piqueux, Arizona State University
Mark Powell, Jet Propulsion Laboratory
Jon Proton, Cornell University
Jim Rice, NASA Goddard Spaceflight Center
Steve Ruff, Arizona State University
Mariek Schmitt, Smithsonian Institution
Michael Sims, NASA Ames Research Center
J. R. Skok, Cornell University
Pamela Smith, Cornell University
Larry Soderblom, US Geological Survey
Nicole Spanovich, Jet Propulsion Laboratory
Steve Squyres, Cornell University
Bob Sucharski, US Geological Survey
Rob Sullivan, Cornell University
Dale Theiling, Cornell University
Ashitey Trebi-Ollennu, Jet Propulsion Laboratory
Roxana Wales, Google
Alian Wang, Washington University, St. Louis
Lorenz Wendt, Freie Universität Berlin
Don Wilhelms, independent scientist
Sandra Wiseman, Washington University, St. Louis
Bo Wu, Ohio State University
Lin Yan, Ohio State University
Aileen Yingst, Participating Scientist

Site Visits

Arizona State University, Phoenix, AZ
Cornell University Astronomy Department, Ithaca, NY
Honeybee Robotics, New York, NY

Jet Propulsion Laboratory, Pasadena, CA

NASA Ames Research Center, Mountain View, CA

Ohio State University, Columbus, OH

SETI, Mountain View, CA

Smithsonian Institution, Washington, DC

Space Science Institute, Boulder, CO

US Geological Survey, Flagstaff, AZ

Washington University, St. Louis, MO

Meetings Attended

Science Operations Working Group meetings (observed daily from *Spirit* SOWGs sol 945 to sol 1362, and *Opportunity* sol 924 to sol 1338)

End of Sol meetings, August 30, 2006, to November 19, 2007

MER Science Team Meetings, February 2007, July 2007, January 2008, January 2009, July 2011

Seventh International Conference on Mars, July 2007

American Geophysical Union Conference, December 2007 and December 2008

Lunar and Planetary Society Conference, March 2009

Mars Exploration Program Advisory Group meeting, July 2007 and September 2008

Mars Science Laboratory Landing Site Selection meeting, September 2008

Spirit rover funeral: team-only celebration and NASA special event

APPENDIX C
Abbreviations and Definitions

AI: Artificial intelligence.

AO: Announcement of Opportunity: An official call from
            NASA that invites proposals for new missions.

APXS: Alpha Particle X-ray Spectrometer: an Athena science
            instrument on the rovers.

ARC-GIS: Software used in Geographic Information Systems for
            producing maps and integrating geographical datasets.

ATC: Approximate True Color: an algorithm for producing im-
            ages that approximate the color sensitivity of the human eye.

Athena: The project name for the suite of interrelated instruments
            that the rovers carry as their "science payload." The sci-
            ence team is also called the Athena Science Team.

CCD: Charge-coupled device, a digital photographic plate.

DEM: Digital elevation map, a topographical mesh of the local
            terrain generated from image data.

ENVI: Image-processing software made by Exelis, often used in
            Planetary Science.

EOS: End of Sol: a weekly meeting at which scientists discuss their
            ongoing scientific work and Long Term Planning issues.

fMRI: Functional magnetic resonance imaging, a type of brain scanning
               produces maps of brain activity.

GIS: Geographic (or Geographical) Information System. Assists in plot-
               ting map and terrain data and overlapping datasets.

Hazcam: Hazard Avoidance Cameras: two pairs of cameras with fish-eye optics
               mounted on the front and rear of the rovers under the deck looking
               over the wheels.

HiRISE: High-Resolution Imaging Science Experiment: color camera on
               board the Mars Reconnaissance Orbiter.

HRSC: High-Resolution Science Camera onboard Europe's Mars Express
               Orbiter.

IDD: Instrument Deployment Device: a robotic arm.

IDL: Interactive Data Language, a commercial software by Exelis fre-
               quently used by astronomers for image processing.

IOF: "I over F," a radiance factor computed during image calibration.

ISIS: Image-processing software suite by the US Geological Survey for
               Planetary Science processing.

ITAR: International Traffic in Arms Regulations: the legal restrictions on
               foreign nationals involved in American space missions.

JPL: The Jet Propulsion Laboratory in Pasadena, CA: Part of the Califor-
               nia Technical Institute (Caltech), JPL is a NASA contractor for most
               robotic missions, such as the rovers. JPL engineers tested, built, and
               now operate both rovers on Mars.

KOP: Keeper of the Plan.

Long baseline stereo (or wide baseline stereo):   Producing topographical data
               by driving the rover several meters between images. This increases
               the distance ("baseline") between the stereo images and therefore
               captures more topographical detail in the resulting model.

LTP: Long Term Planning: the activity of producing "strategic" plans
               for the rovers—longer duration goals for driving or science. LTP
               discussions occur during End of Sol meetings. The LTP Lead is the
               scientist in charge of managing these discussions and keeping the
               big picture in mind during the immediacy of "tactical," or day-to-day,
               operations. This position rotates every few weeks among a group of
               mission scientists.

Maestro or SAP: The rovers' science activity planning software. Allows the team
               to keep track of planned scientific observations alongside engineering
               operations over the course of the day on Mars.

MEP: Mars Exploration Program.

MEPAG: Mars Exploration Program Advisory Group.

MER: Mars Exploration Rover mission. Often used interchangeably with
               the rovers' nicknames: *Spirit* is officially MER-A, and *Opportunity*
               is MER-B.

MI: Microscopic Imager: an Athena science instrument on the rovers.

MiniTES: Miniature Thermal Emission Spectrometer: an Athena science
               instrument on the rovers. Modeled on TES and Themis, the Thermal
               Emission Spectrometers in orbit around Mars on *Odyssey* and the
               Mars Global Surveyor orbiter respectively.

MOC: Mars Orbiter Camera: built by Malin Space Science Systems, on
               board the Mars Global Surveyor orbiter; an earlier version was lost
               with the Mars Observer spacecraft.

MOLA: Mars Orbiter Laser Altimeter: an instrument on board the Mars
               Global Surveyor that used a laser sensor to determine the topography
               of Martian terrain. The colorful MOLA map of Mars is considered
               the standard topographical projection for Mars: one MER scientist
               called it "the control for the planet."

Mössbauer: The Mössbauer spectrometer, an Athena science instrument on the
               rovers.

MRO: Mars Reconnaissance Orbiter: a NASA vehicle in orbit around Mars
               from November 2006.

MSL: Mars Science Laboratory, the next-generation rover, nicknamed
               Curiosity. Originally planned to launch in 2009, rescheduled to 2011,
               landing on Mars in 2012.

NASA: National Aeronautics and Space Administration: the sponsoring
               agency for the Mars Exploration Rover project.

Navcam: The Navigation Cameras, stereo black-and-white engineering cam-
               eras mounted on rovers, slightly inset from the Pancams.

OMEGA: An orbital spectrometer built by a French team, aboard the European
               Space Agency's Mars Express orbiter.

Pancam: Panoramic Cameras, the "science cameras" on the Athena science
               payload, providing stereo and color imaging for the mission.

Participating Scientist: A scientist that NASA selected and funded to participate
               directly on the Rover mission, including requesting, planning, and
               analyzing rover observations and, sometimes, supporting a lab of
               graduate students or staff researchers working on the mission.

PCC: Pancam Calibration Crew.

PDL: Payload Downlink Lead: person in charge of monitoring instrument
            health and recent communication activities from the spacecraft. This
            position rotates among a group of scientists.

PI: Principal Investigator: the lead scientist on the Mars Rover team. Un-
            like larger missions, there is only one PI on the Rover mission.

PUL: Payload Uplink Lead: person in charge of compiling commands for
            the rovers' upcoming operations daily. In the case of the Pancams,
            RAT, and MI, the position is occupied by one of a few specially
            trained PULs; PULs for other instruments rotate among engineers,
            scientists, and graduate students on the team.

RAD: A radiance constant computed during image calibration.

RAT: Rock Abrasion Tool: a grinding tool that can brush or grind away
            outer "rinds" of rocks; part of the Athena science suite on board the
            rovers.

RP: Rover Planner: specialist engineer responsible for driving the rovers.

Sol: A Martian solar day, 24 hours, 39 minutes, and 35 seconds long.
            Also used to abbreviate the day of the mission: sol 1500 is the rover's
            1,500th day on Mars. Each rover landing counts as sol 1 for that
            rover.

SOWG: Science and Operations Working Group (pronounced either "sŏg"
            or "ess oh double-u gee"), the daily tactical meeting of scientists and
            engineers in which a plan is agreed on for the next day's operations.

STG: Science Theme Group, a loose cluster of Participating Scientists on
            the mission who determine a shared set of problems or interests in a
            domain such as atmospheric sciences or geochemistry, then outline
            observations across multiple instruments to solve these problems.
            The STG sends a representative to the SOWG meeting (a rotating
            role) to advocate for these observations.

Team Meeting: Face-to-face meetings among the science team members when
            scientists meet for two days to discuss their ongoing scientific results
            and Long Term Planning strategies. During my fieldwork this took
            place approximately once a year, then transitioned to once every two
            years to accommodate budget cuts.

TES: Thermal Emission Spectrometer, in orbit on the Mars Global Sur-
            veyor (lost November 2006). TES is the orbital version of the rovers'
            MiniTES.

THEMIS: Thermal Emission Imaging System, infrared spectrometer in orbit
            on the Mars *Odyssey* orbiter. Made by the same group of scientists

who made TES and MiniTES. THEMIS infrared data is used as the baseline for most geological maps of Mars.

USGS: United States Geological Survey.

VIZ: Software visualization suite developed at NASA Ames Research Center, specializing in three-dimensional modeling.

NOTES

Introduction

1. Lowell, *Mars*; see also Lane, "Geographers of Mars"; Lane, *Geographies of Mars*.

2. Galison, "Judgment against Objectivity"; Tucker, *Nature Exposed*.

3. These questions are especially salient in the history of representation in scientific practice. Scholars have shown that rather than resolving questions of visual authenticity, new representational technologies such as photographic cameras instead open the question for debate. See in particular Galison, "Judgment against Objectivity"; Daston and Galison, *Objectivity*; Tucker, "Photography as Witness."

4. Hooke, *Micrographia*, preface. On naturalism as a visual convention, see Kemp, "Taking It on Trust."

5. Related to the "principles of inclusion and exclusion" in images discussed by Gordon Fyfe and John Law, "On the Invisibility of the Visual," 1.

6. Hanson, *Patterns of Discovery*, 7.

7. Goodwin, "Professional Vision." Compare this with the skilled seeing John Law and Michael Lynch describe in their study of birdwatching ("Lists, Field Guides"), in which not only recognizing particular birds but also relating them to their representations in field guides requires acquiring a visual skill.

8. A recent stream of scholarship in science and technology studies has described the embodied practices of visualization, which I engage in more detail in chapter 6. See Alač, *Handling Digital Brains*; Myers, "Molecular Embodiments"; Radder, *World Observed*; Prentice, *Bodies of Information*.

9. As Klaus Amman and Karin Knorr-Cetina remind us, "*seeing is work*" ("Fixation," 90). Describing how biologists interpret gels and autoradiograph films, the two researchers show that it is only through focused conversation and interaction about an image that scientists come to see the information hidden therein. Michael Lynch analyzes how scientific seeing is inextricably linked to representational techniques such as selecting or mathematizing particular elements in the visual field (Lynch, "Externalized Retina") or disciplining the object of analysis into compliance with visual modes (Lynch, "Discipline"). Bruno Latour's discussion of *inscriptions* also notes the importance of "thinking with eyes and hands" when scientists engage in what he terms "the transformation of rats and chemicals into paper" (Latour, "Visualization and Cognition"; cf. Latour and Woolgar, *Laboratory Life*).

10. For example, Alač, *Handling Digital Brains*; Beaulieu, "Images"; Edwards, *Vast Machine*; Joyce, "From Numbers to Pictures."

11. Knorr-Cetina and Amman, "Image Dissection," 280. The illustration is not innocent: "Analyzability . . . is built into the record from the beginning." Amman and Knorr-Cetina, "Fixation," 107. This analytical lens brings our attention to what Coulter and Parsons call the praxiology of perception: those practical activities, forms of talk, interaction, imaging conventions, and instrumental techniques that scientists use to make sense of visual materials. Coulter and Parsons, "Praxiology of Perception," 252.

12. Archival documents reveal that NASA had explored the possibility of sending robotic vehicles to Mars beginning in the 1960s. In one iteration of the Viking missions of the 1970s, a follow-up *Viking* 3 and 4 were meant to have roving capabilities.

13. On this topic see McCurdy, *Faster, Better, Cheaper*; Kaminski, "Faster, Better, Cheaper."

14. Hubbard, Naderi, and Garvin, "Following the Water."

15. On the choice of astronauts in the US space program and the shaping of spacecraft systems, see Mindell, *Digital Apollo*; see also Gerovitch, "'New Soviet Man'" for a comparative approach.

16. Because the Hazcam and Navcam images are used to make decisions about where and how the rovers can drive, they are considered the rovers' "engineering cameras" and were developed as part of their robotic hardware at the NASA Jet Propulsion Laboratory. The Pancams and Microscopic Imager are considered part of the rovers' scientific suite of instruments (the Athena payload), were developed with NASA funding but not based at a NASA center, and are usually commanded to take images related to specific experiments or scientific investigations.

17. In contrast, the European Space Agency releases AOs that call for fully funded contributions from its member countries, with ESA itself providing coordination but limited operating costs. On ESA, see Zabusky, *Launching Europe*.

18. The division between "science" and "operations" is a consistent feature of spacecraft design in robotic space exploration. Establishing and maintaining these categories and their associated social relations does important work for such teams: it can elucidate where particular lines of funding should be directed or articulate which aspects of the mission are state secrets versus information that can be shared with international partners (more on this below).

19. For an examination of the primary, colocated phase of mission operations at JPL, see Clancey, *Working on Mars*; Mirmalek, "Solar Discrepancies"; Mirmalek, "Working Time on Mars"; Tollinger, Schunn, and Vera, "What Happens"; Wales, Bass, and Shalin, "Requesting Distant Robotic Action."

20. Zara Mirmalek ("Solar Discrepancies," "Working Time on Mars") conducted a detailed study of coordinating work across these time scales during primary operations.

21. Goffman, *Interaction Ritual*.

22. Social organization is not equivalent to social order. Locally accountable actions and interactions that make up the social order may or may not correspond to organizational form. For example, there may be formal rules of interaction dictated by hierarchy or organizational form, while at the same time informal rules guide interactions and produce order on the ground. However, the organizational attributes of the Rover mission serve as an important narrative, grounding ritual interactions and accounting for activity. This ties both formal and informal aspects of social order together in practice in this institutional context.

23. This does not mean there is no hierarchy on the Rover mission. A single Principal Investigator leads the team: he is a charismatic personality who, through his work with the Rover team, has become a well-known figure in NASA science and politics. (Note that my use of "charisma" is both an actor's category and an analytical one: see Weber, *Theory of Social and Economic Organization*.) The mission combines engineers embedded in JPL's matrix structure, scientists at all stages of their careers from mission veterans to undergraduate students, participants housed at institutions of various degrees of status, and of course NASA's own bureaucratic hierarchical structure. Given this heterogeneous group of participants, the mission's consistent attention to collectivity and consensus must be understood as a continuing social achievement.

24. On organizational attributes of laboratory practice, see Knorr-Cetina, *Epistemic Cultures*; Shrum, Genuth, and Chompalov, *Structures of Scientific Collaboration*; Salonius, "Social Organization of Work"; Turner, "Where the Counterculture Met the New Economy"; Doing, "'Lab Hands' and the 'Scarlet O.'"

25. As I will show, much of the work of the mission is oriented toward this narrative. It provides an important organizational rationale; the central members' account through which all decisions must be explained; and a resource for negotiation, group management, and decision making. In Meyer and Rowan's terms, it is the most powerful story within the organization that is enacted through ritual and reinforced for individuals as the institutional imperative. Meyer and Rowan, "Institutionalized Organizations."

26. See Perlow and Repenning, "Dynamics"; Polletta, *Freedom*.

27. A parallel here is in Ed Hutchins's work on distributed cognition. In his work on ship navigation, Hutchins draws our attention to the organizational aspect of navigation. Knowing where the ship is at any given time is not the responsibility of any single individual. Instead, this knowledge is always socially achieved through a distributed organization of personnel who deploy individual ways of knowing, seeing, and representing in an organizational and communicative context to resolve that single problem. Navigation thus not only depends on singular instruments, it also depends on communicative practices, institutional roles, cultural differences, and distribution of individuals across a site, even in the context of a single entity like a ship. Hutchins, *Cognition in the Wild*.

28. Scott, *Seeing like a State*.

29. Liz, personal conversation, February 6, 2008.

30. Laboratory ethnography is an established method in science and technology studies. For formative examples, see Knorr-Cetina, *Epistemic Cultures*; Latour and Woolgar, *Laboratory Life*; Lynch, *Art and Artifact*; Traweek, *Beamtimes and Lifetimes*.

31. Tracing individuals' activities with images can be difficult when they are sitting alone at their screens. How are we to identify their assumptions and rituals, their embodied situations, their organizational locations, and their everyday practices then, when everything appears to simply be individual cognitive work? To surmount this problem, certain scholars in science

studies have taken to studying scientists at work with images in teaching or training settings, when they must verbalize the tacit assumptions of their field (see Alač, *Handling Digital Brains*; Prentice, "Drilling Surgeons"; Kaiser, *Pedagogy*). On the Rover team, I took advantage of a different facet of mission work: teleconferences. Because team members are distributed across the United States (and some of Europe), they meet using teleconference facilities, with very limited video. This means that even the most visual of interactions had to be made explicit through language, annotations, and a range of other resources. Seeing moved from being something that happened in the privacy of a scientist's computer terminal, retina, or brain to an activity that was practical, interactive, and observable (or in Garfinkel's terms, "observable-reportable" [*Studies in Ethnomethodology*]).

32. This multisited approach (Marcus, "Ethnography") honed my attention to the mission's many internal partial perspectives (Haraway, *Simians, Cyborgs, and Women*) and enabled a more holistic view of the mission.

33. The rare exception is when the identity or role of a public figure is crucial to understanding a particular decision or representation.

34. Harry Collins and Robert Evans ("Third Wave") describe this as "interactional expertise": the ethnographer's ability to understand and interact knowledgeably with a group of scientists without becoming a scientist or practitioner oneself.

35. Ethnomethodology examines the construction of social order as established through everyday, practical activities, accounts, and interactions. My attention to the material, practical, quotidian, and interactional elements of making knowledge about Mars while crafting local social order is indebted to this approach. See Garfinkel, *Studies in Ethnomethodology*; Garfinkel, *Ethnomethodology's Program*; Lynch, *Scientific Practice*.

36. The science meetings and downlink-related science activities I witnessed were not subject to technical restrictions. However, I did not have access to e-mails distributed on the Rover Listserv or to document-sharing sites related to uplink activities. I was not permitted to view the programming of rover operations through the team's software tools, and I did not attend any meetings where technical details or sequencing were discussed; nor did I witness any of the code or technical side of production and uplink to the rover. Backstage chatter that usually occurs over teleconference lines after the open meetings was also off-limits owing to implied virtual presence in the engineers' workroom. My participants were clearly informed of my status as a foreign national and their responsibilities to uphold ITAR before I conducted any interviews or observations of their scientific work. Any discussion in this book that touches on the technical side of the rovers is anecdotally derived and technically nonspecific or is published and therefore in the public domain. When I refer to rover operations, therefore, this does not include any technical details of rover operations that might constitute a security violation. For another example of an ethnography conducted under conditions of limited access, see Hugh Gusterson's study of the Lawrence Livermore National Laboratory, *Nuclear Rites*.

37. This may appear irksome to scholars in science and technology studies, who are accustomed to discussing the social, scientific, and technical as intrinsically interrelated and indistinguishable. However, this distinction remains in my work as an artifact of my access to the field site. Further, "science" and "operations" are some of the strongest actor's categories on the mission. They are distinctions made and enforced by the team for the purposes of distributing responsibilities for the spacecraft in the management of rover resources, maintaining professional identities on a multidisciplinary and multi-institutional team, managing consensus and interdisciplinary communication, and maintaining compliance with federal regulations. My own atten-

tion to these terms' deployment in the field therefore does not imply illegal access to restricted technical details, on the one hand, or ignorance of core science studies concepts, on the other, but is discussed in terms of the sociological work of this actor's distinction in context.

38. Knorr-Cetina and Bruegger, "Market as an Object of Attachment."

39. Note that when I mention the rovers, I deploy them as the team does: as a narrative device, not as narrating individuals (as in, for example, Latour, *Aramis*). I describe this in more detail in chapter 6.

40. This organizational aspect is frequently overlooked in actor network or ontological approaches to science studies, which assume symmetry between all nodes of a sociotechnical network, whether human or machine. But even actor networks have a topology, and local ontologies developed under these arrangements reflect organizational hierarchies. For example, the Mars rovers might be considered "distributed subjects," assembled from a heterogeneous collection of individuals and machinery on Earth (in the style of Hélène Mialet's *Hawking Incorporated*), but these aspects are always "distributed" according to a particular organizational order. It could also be instructive to consider human-machine interactions in the style of a multispecies or cyborg perspective (Haraway, *When Species Meet*). Such an approach must consider the local hierarchies that imbue the human-robot encounter and how different humans (and robots) perform different organizational positions at their interface.

Chapter One

1. This is a common opening statement at SOWG proceedings, indicating that the meeting is about to begin. Descriptions of and quotations from SOWG interactions in this chapter were recorded and transcribed during meetings observed between 2006 and 2008. For more on the SOWG, see Tollinger, Schunn, and Vera, "What Changes"; Cheng et al., "Opposite Ends of the Spectrum."

2. The three participating institutions with video feeds are located with the Principal Investigator, the Deputy Principal Investigator, and the Jet Propulsion Laboratory. During my site visits, I noted other polycom units, none in use. Most participants phoned in to the telephone line from their individual desks, including those who had colleagues down the hall participating in the same mission. Although each participant in principle has the same level of access to mission materials through networked sites and video links, in practice they have varying degrees of access, types of documents (static or refreshing), and people with them as they work together on this collaborative and virtual project.

3. Pronounced sŏg.

4. Rover artificial intelligence is limited to autonavigation around features in the Martian terrain that the robots might judge insurmountable in situ. This has been subject to upgrade over the course of the mission as new software is uploaded to the rovers—as team members put it, "our rovers are getting smarter." The emphasis on human actors is particularly important to William Clancey ("Clear Speaking about Machines"), who argues that calling the rovers "robotic geologists" obscures the human element of the mission.

5. Interview, Mark, February 15, 2007.

6. The relation between ritual and social order has long been of interest to sociologists. Émile Durkheim described such order as arising from the performance of religious rites and the social division of labor (Durkheim, *Division of Labor in Society*; *Elementary Forms of the Religious Life*). Erving Goffman (*Interaction Ritual*) describes such order as arising from face-to-face ritual

interactions, a point later articulated in detail by Randall Collins (*Interaction Ritual Chains*). Alternatively, Harold Garfinkel's program of ethnomethodology focuses on the routine grounds of practical action, local sense making, and actor's accounts (see Garfinkel, *Ethnomethodology's Program*, 92–93). Such authors present different approaches that cannot be seamlessly combined. I am indebted to Durkheim and to Collins for the theme of social solidarity arising from these types of interactions, although I resist Durkheim's association of these activities with primitive social groups. However, I adhere to Goffman's and Garfinkel's observations that social order is produced not through macrolevel structures, but through moments of face-to-face encounter, talk, and interaction. This requires considering how individuals produce social order through talk and practice and noting what resources are available to them to do so. In line with Goffman and Collins, then, I describe team rituals; with Garfinkel (*Studies in Ethnomethodology*) I focus on routine activities, often visible through moments of breach, for their role in producing social orderings and group membership.

7. Because activities at the SOWG meeting constitute the everyday work of conducting science on Mars, image planning in the SOWG presents an interesting site for the exploration of scientific representation as what Garfinkel would call "ordinary action." On the application of ethnomethodological techniques to science, see Lynch, *Scientific Practice and Ordinary Action*.

8. Known as the Callas Palace, named for the project manager who oversaw its construction to these specifications, the room sported light wooden tables arranged in a U, with multiple screens at the front of the room and a model rover (nicknamed *Buddy*) at the center. This room was in operation until 2011, when it was dismantled to make way for *Curiosity*'s operations.

9. One scientist I spoke to, James, traced the origins of this structure to the testing phase of rover operations, in which it became clear that for the team and the rover to operate together successfully, "you need some degree of organization, and there are crucial positions that need to have folks with skills associated with them." This required "the right partitioning of assignments" among people on the team. Interview, James, June 21, 2007.

10. Most of these instrument liaisons combine both downlink and uplink responsibilities; only the Pancam has distinct liaisons, in part owing to the volume and complexity of data that must be managed.

11. By the time of my fieldwork, the chairs behind these placards were usually empty, their occupants dialing in to the meeting from across the country. The placards stand as reminders of the virtual participants at the table.

12. The Principal Investigator chooses the Chairs for their ability to manage complex negotiations without disenfranchising team members. The roster of Chairs includes members of the Rover team who have had experience with leading missions, but it also includes scientists who are new to mission participation. The role has been described to me in various ways, but all team members essentially concur with scientist Alexa's observation that "not everybody's suited to be a SOWG Chair." The managerial aspect of the role appeals in particular to the younger scientists who are invited to assume the position: as Alexa explained, "I'm ambitious. . . . I want to run my own mission someday." She admitted that managing so many different (and often famous) personalities, having to make difficult calls and build consensus, and making a decision about what the rover should do was daunting at first, but it was also exciting and felt like good "on-the-job training" for the next stage of a successful planetary scientist's career (Alexa, personal conversation, September 16, 2008). This is consistent with the tutelage approach in collectivist organizations as identified by Francisca Polletta (*Freedom Is an Endless Meeting*), where organizers see

their role as developing the talents and skills of their recruits, grooming certain individuals for leadership roles.

13. Interview, James, June 21, 2007.

14. The Chair's pronouncement was met with agreement by others in the room, who noted, off the microphone, "That's exactly what I was going to say," "It has nothing to do with today's tactical plan," and "I can't figure out why after 1100 sols, people still haven't figured out that you don't talk about strategic issues in tactical meetings" (*Opportunity* SOWG, sol 1107, March 5, 2007).

15. Some of the roles are jobs for specialist technicians, but many (such as PDLs, KOPs, or LTP Leads) are filled by Participating Scientists who are not committed to a particular instrument's operation. The scientists are loosely grouped into science theme groups (STGs) sharing common research questions, such that they may together craft a list of observations that their discipline finds important or salient. Each STG designates a member to attend each SOWG to represent the group's interests, concerns, or requests. Participation at the meeting was not restricted to these STG Leads, nor was priority accorded to STG observations in determining a day's plan. During my observations of the team these groups did not function as competing political units but rather were visible as a roster of Leads who could be called on to speak up for one or another scientific perspective during a meeting.

16. These positions are normally accomplished remotely: that is, the Documentarian, KOP, Chair, and LTP lead are rarely, if ever, in the same room at the same time. Compare with SOWGs described during the colocated primary mission in Tollinger, Schunn, and Vera, "What Changes."

17. Team Meeting, January 13, 2009.

18. Interview, James, June 21, 2007.

19. Interview, James, June 21, 2007.

20. *Opportunity* SOWG, sol 933, September 13, 2006.

21. *Opportunity* SOWG, sol 1102, March 1, 2007.

22. I discuss naming conventions in chapter 4.

23. I return to this point with its embodied implications in chapter 6. On other formulations of place and membership, see Schegloff, "Notes on a Conversational Practice"; also Sacks, *Lectures on Conversation*, especially II.3 and III.8.

24. LTP Leads typically occupy their role in shifts of about two or three weeks at a time to maintain continuity between strategic discussions and tactical implementation.

25. *Spirit* SOWG, sols 1128–29, March 5, 2007. Although updated daily, LTP reports typically preserve several key slides embedded in the presentation for days at a time.

26. *Spirit* and *Opportunity* are on different sides of the planet, and each has its own SOWG meeting. Their PowerPoint presentations have stabilized into two slightly different documents. On *Spirit*, long-term planning objectives are more often displayed in a graph or table, and there is always a graph of current data volume and power; on *Opportunity* these are usually displayed as units on a different slide. Rover team members account for these differences as reflections of the differences between the two rovers and their teams, who are often said to have different personalities, as I will describe in chapter 6.

27. Called tau, this is a measurement of how much dust is in the atmosphere, what atmospheric scientists call optical depth. On the Rover mission this is measured by taking pictures of the sun with the panoramic cameras. Since scientists know how bright the sun ought to look in these photographs, they compare how much its brightness is reduced to characterize the dust in

the atmosphere and estimate solar power levels. Tau graphs are presented at the outset of every SOWG meeting.

28. The solar-paneled vehicles may have lots of energy to power observations during summer days when the sun is high, but during the winter or during a dust storm these levels may become dangerously low. During the dust storm of 2007, for example, *Spirit* had only enough energy to transmit a single beep to Earth a few times a week and could not even take pictures of the sun for tau measurements, since the sky was so obscured by dust that its solar detection program could not locate the sun.

29. This report and a few others from engineering personnel were added to the SOWG roster of reports during my observations of the mission. This was suggested by a JPL engineer as a good way to maintain what he called "cohesiveness" between the science and operations sides of the mission: by keeping all team members—not just those directly responsible for producing rover commands—informed about the status of their vehicle.

30. For example, the skeleton for the day may show that the rover wakes up at 10:00 a.m. (Mars time) and has two hours available "for science" before it must check in with its relay orbiter, *Odyssey*, flying by overhead; it then must take a "nap" to recharge, "wakes up" at 2:00 p.m. and has enough energy to drive twenty meters to its next target site, leaving only twenty minutes "for science" at the end of the drive before it has to "go to sleep" overnight at 4:30 p.m.

31. Because team members rotate regularly through these roles, this presents an opportunity for fresh eyes on a problem or time to resolve any conflicts between team members as they arise during planning, before they become personal.

32. Garfinkel (*Studies in Ethnomethodology*) articulates "accounting" as actors' sense-making practices in the context of ordinary activity. The SOWG meeting can be seen as an elaborate networked social setting for accounting for each rover's activities. Within the space of "minding the bit bucket," then, sensible requests from team members are those that satisfy the changing parameters of rover resources.

33. *Opportunity*, SOWG, sols 947–49, September 22, 2006.

34. *Spirit* SOWG, sol 977, October 2, 2006.

35. *Opportunity* SOWG, sols 1100–1101, February 26, 2007.

36. *Opportunity* SOWG, sols 954–56, September 29, 2006. Note that these requests do not come out of the blue: they usually have a history in the weekly science meetings, discussed in chapters 4 and 5.

37. *Opportunity* SOWG, sol 1010, November 26, 2006. The rovers do not have movie cameras on board, but scientists may request several sequential frames of the same observation to assemble into a movie on Earth and thereby identify drift direction of Martian clouds or the formation of dust devils.

38. *Opportunity* SOWG, sol 1102, February 28, 2007.

39. *Opportunity* SOWG, sol 958, October 4, 2006.

40. *Spirit* SOWG, sols 1128–29, March 5, 2007.

41. *Spirit* SOWG, sols 1118–20 February 23, 2007.

42. These are "bookkept" by the KOP to support the drive described on p. 35: Pancams within the drive and ultimate and penultimate Hazcam images. For more details on driving images, see Maki et al., "Operation and Performance of the Mars Exploration Rover Imaging System."

43. *Opportunity* SOWG, sol 953, September 28, 2006.

44. *Opportunity* SOWG, sol 1075, January 31, 2007.

45. The Rover mission is deeply attuned to this division, and much of the mission's organizational structure and narratives is geared toward actively bridging this divide. As scientists circulate through light engineering roles such as Downlink Lead, they get a feeling for the spacecraft and forge friendships with the engineers on the line. Engineers attend science meetings regularly, and twice I witnessed scientists fly to JPL to work alongside engineers to help solve the problem when *Opportunity*'s and then *Spirit*'s wheels were lodged in soil.

46. *Opportunity* SOWG, sols 943–44, September 18, 2006.

47. *Opportunity* SOWG, sol 953, September 26, 2006.

48. *Spirit* SOWG, sol 1102, February 7, 2007.

49. *Opportunity* SOWG, sols 1063–65, January 19, 2007. At first glance it is tempting to view such interruptions as abuse of the SOWG Chairs' power: they can even convince the final arbitrators of the plan's viability (the Mission Manager and Long Term Planner) that their observation should occur regardless of duration and bits consumption. However, other team members may request similar observations, although they usually require the Chair's buy-in to make them happen. Still, scientists on the team do not consider these breaches of the regular rules to be problematic, disruptive, or outside the Chair's authority. One scientist shrugged when such an observation went through, saying, "She's the Chair. If that's what she wants to do, that's fine with me." Such moments point to the continued importance of leadership even on a flattened-hierarchical collectivist team.

50. The development of the field sciences in early twentieth-century America and their relation to laboratory-based models of scientific work is eloquently explored in Kohler, *Landscape and Labscapes*. The rovers present a challenging case, since the field is simultaneously physically remote and virtually present, the laboratory is often located "inside" a computer, and the scientists who populate the mission hail from both lab (chemistry) and field (geology) disciplines. I discuss this further in chapter 7.

51. LTP report slide, *Opportunity* SOWG, sols 943–44, September 18, 2006.

52. Targets of opportunity are defined by the team as potential sites of interest along a drive path that deserve closer inspection but usually become apparent only at the last minute (e.g., during a drive).

53. Landis, "Some MER Terminology." As scientists Roger and William each explained to me, the term harked back to a moment during operational readiness tests on Earth before the rovers were built. One group was placed in a bunker with instructions to operate the model rover at a distance without knowing where it was; another group was located in the field with the rover to see what the bunker team would do. To test the group in the bunker, the field group placed a fossil just behind the rover. To their amusement, the group in the bunker never discovered the fossil because they kept imaging and driving forward instead of looking around more frequently.

54. *Opportunity* SOWG, sol 958, October 4, 2006.

55. As this postcard developed into a high-resolution black-and-white panorama, it came to be known as the "Ansel Adams Pan," taking on a sense of the untouched natural landscapes that enticed adventurous Americans to the frontiers of their country. I will discuss this element of Martian imaging further in chapter 8.

56. Liz, personal conversation.

57. *Opportunity* SOWG, sols 1191–93, May 30, 2007.

58. *Spirit* SOWG, sol 953, September 8, 2006.

59. *Spirit* SOWG, sols 1138–40, March 15, 2007.

60. *Opportunity* SOWG, sol 952, September 27, 2006.

61. *Spirit* SOWG, sol 1164, April 11, 2007.

62. *Spirit* SOWG, sol 1034, November 27, 2006.

63. *Spirit* SOWG, sols 1193–95, May 11, 2007.

64. *Opportunity*, SOWG, sol 1033, December 19, 2006. The TAPSIE (Tactical Activity Planner/ Sequence Integration Engineer) is responsible for producing the backbone of the day's plan.

65. Randall Collins (*Interaction Ritual Chains*) argues for a relation between those interaction rituals that produce shared affective states in participants and the group's social solidarity. Whether or not individuals' stated "happiness" corresponds to an actual affective state, stating and restating satisfaction in these terms certainly builds up the team's emotional energy at a distance and produces a particular solidarity among the group members. In a ritual performance reminiscent of Durkheim's effervescent assembly (*Elementary Forms of the Religious Life*), this ritual and affective statement builds and sustains collective engagement. Counter to Collins's claim that such practices require physical copresence, the rover example demonstrates how software suites and digital copresence may indeed satisfy in producing solidarity.

66. Affective talk and action continue outside this ritual response pair to maintain team members' emotional energy and, concomitantly, their commitment to the organization. Even during a stressful period of deciding where *Spirit* should drive at Home Plate, a team member characterized his colleagues as "driving around and having a great time." When charged with a difficult piece of rover planning requiring custom commands, I witnessed a team member walk away from the table at the end of the meeting singing, "I get to plan a custom sequence, do-de-do-de-do!" Exchanges about negotiating scientific observations frequently display jovial framing, poking fun at exactly the combination of serendipity and flexibility that characterizes the daily situated planning process. Laughter often rings out on the line, scientists whoop and cheer as new images come down from the rovers, people promise each other "beer bets" over whether a "perfect" observation will materialize, and I have witnessed more than one team member regularly leave the SOWG meeting saying, "This is one kickass mission!" I will return to the role of affect in team solidarity in chapter 6.

67. *Opportunity* SOWG, sols 954–56, September 29, 2006.

68. Team Meeting, July 7, 2007.

69. End of Sol, September 13, 2007.

70. Garfinkel's (*Studies in Ethnomethodology*) breaching experiments, for example, attempted to force altered moments in patterns of exchange in order to probe the underlying rules to everyday sense making. As group members work to repair the breach, they often express the very tacit underlying rules of social order that were breached in the first place. These moments of exception therefore do not disprove the underlying rule but rather expose it.

71. Field notes, *Spirit* SOWG, sols 1148–49, March 26, 2007. Matters were not helped by a new phone system implemented in the wake of a security breach. Participants were for the first time asked to state their names as soon as they dialed in. Thus the meeting was interrupted repeatedly as the Mission Manager tried to figure out who was on the line to satisfy local security requirements.

72. The Principal Investigator even asked the sociologist present at the meeting for "any idea of what went wrong here." S. Squyres to J. Vertesi, e-mail correspondence, March 27, 2007.

73. *Spirit* SOWG, sols 1150–51, March 28, 2007.

74. It also represents a potential miscommunication between a spectroscopist (the requesting scientist) and a geomorphologist (the Chair), another disciplinary tension within planetary science. I will describe this particular tension and its local resolution in more detail in chapter 5.

75. Field notes, *Spirit* SOWG, sols 1150–51, March 28, 2007.

76. Personal conversation, March 28, 2007.

77. Lewis et al., "Structure and Stratigraphy of Home Plate."

78. *Opportunity* SOWG, sol 953, September 28, 2006.

79. This "glory Pancam," or "Lion King Pan," as team members called it during the planning process, was the topic of much speculation among scientists over which view of the crater would have "more pizzazz." When the rover arrived at the chosen location on sol 952, it acquired a Navcam panorama. The Navcam images were used to plan a Pancam four-filter color panorama to produce an approximate true color image for public release: a process I will describe in chapter 8. On "chains of inscriptions," see Latour, "The 'Pedofil' of Boa Vista."

Chapter Two

1. Squyres, *Roving Mars*, 168. In this chapter I focus on the Pancams, but the instrument leads each maintain their own calibration routines at their own institutions.

2. On the center of calculation, see Latour, "Visualization and Cognition." Of special concern was whether the instrument would behave the same way abroad as it did at home (Delbourgo and Dew, *Science and Empire*): whether crafting timepieces to establish longitude at sea, disciplining individual scientists to eliminate the "personal equation" from their astronomical observations (Schaffer, "Astronomers Mark Time"), or formulating the standards for the meter or the pound (Wise, *Values of Precision*; Star and Lampland, *Standards and Their Stories*). Postcolonial histories and inversions aside, metrology is also a practice concerned with exerting control from center over periphery (Schaffer, "'On Seeing Me Write'").

3. In the context of controversy, studied in the early Empirical Programme of Relativism, critiques about instrumental calibration can take on heightened significance. Harry Collins points to precisely the philosophical problem that the Rover PI outlines above: how to detect something—whether a spectral signature on Mars or a gravity wave—that has never before been detected. Collins (*Changing Order*) has effectively argued that calibration, a "test of a test," can only complete the vicious circle of the *experimenter's regress*. That is, fine-tuning an instrument to produce good results requires acting on a preconceived notion of what good results are, what they ought to look like, and how they can be detected. Trevor Pinch further explores this aspect of calibration in his studies of solar neutrino detectors ("Towards an Analysis of Scientific Observation"). Pinch concluded that successful calibration experiments draw on social and technical resources to limit any challenges of undue "similarity assumptions" between the calibration and the main experiment. However, critics of the sociology of scientific knowledge school have countered this approach by attempting to demonstrate that it is possible to appeal to criteria that are external to the calibration setup to validate results, thus breaking free of Collins's experimenter's regress and the confines of theory-laden experiment (Franklin, "Calibration").

4. The work of calibration takes place away from the collective virtual workspace of the Rover team in a laboratory not connected to the main teleconference lines. This does not mean the work of calibration is invisible: the first image *Spirit* relayed after landing was of the Pancam calibration target. During the hectic first ninety days of the mission, student calibrators worked shifts around the clock to adjust images as soon as they were downlinked from the rovers. Calibration sequences are bookkept in the rover's daily plans during SOWG meetings (as I described in the previous chapter), and Rover scientists must wait for calibrated images to appear on the

shared server before they begin the image-processing analyses that I will describe in the next chapter.

5. A pseudonym.

6. On tacit knowledge, see Polanyi, *Tacit Dimension*; in science, see Collins, *Changing Order*. As I myself became an expert calibrator, I was also called on to train other calibrators and supervise their enculturation into the PCC. And as an early draft of this chapter circulated to the Calibration Crew leads, they set up an informational meeting for new calibrators to assemble the group for additional training. Such opportunities recall David Kaiser's work on pedagogy in the sciences (*Pedagogy and the Practice of Science*); and Morana Alač's studies of brain scans (*Handling Digital Brains*).

7. Alač, *Handling Digital Brains*.

8. Field notes, February 2, 2006.

9. Field notes, February 2, 2006.

10. Many of these students were involved in the camera's construction, programming, and preflight calibration. Preflight calibration is not discussed in this chapter: see Bell et al., "Mars Exploration Rover Athena Panoramic Camera (Pancam) Investigation."

11. Goodwin, "Professional Vision."

12. Personal correspondence, August 8, 2007.

13. Calibration Procedures, step 5.04. Data dropouts occur when, owing to interference, the data stream in an image is interrupted, resulting in a big black square in the image. Interestingly, these are seen not as sources of information about, for example, a problem on board the spacecraft or identifying asteroids, space junk, or cosmic rays, but rather as an obstruction in the data, a problem that must be solved by asking the rover to send the image again. Finally, sometimes "pixels get mixed up" (field notes, March 5, 2006) as a result of compression errors. The calibrators' notes in their reports therefore identify to the team which images to request again on the next transmission.

14. Knowing "what Mars looks like" is important for being able to identify errors, artifacts, or novel phenomena. But identifying what an unexpected value means requires a different kind of expertise. If students find something in their data that they sense is suspicious, they are encouraged to contact senior members of the team or make a note in the log for the mission scientists and programmers to review.

15. Field notes, February 2, 2006.

16. The sundial, one of astronomy's most ancient tools, is decorated with a schematic diagram of Earth's position relative to Mars, the planet's name in several different languages, and images selected from a competition of children's drawings. It thus functioned as publicity for the rovers on Earth as well as a being a device that might instruct future visitors to Mars about Earth's civilization. The solar system diagram regularly frustrates calibrators, who curse at the locations of Earth and Mars in the diagram, since they make tagging caltarget regions especially difficult.

17. Field notes, Astronomy 310, lecture 13, November 22, 2005. So central is the notion of ground truth to remote sensing that rover data are sometimes used to help calibrate orbital data from the Mars Express or Mars Reconnaissance Orbiter. On combining these datasets, see chapter 7.

18. Making judgments of similarity and difference between deployment sites is an essential aspect of instrumental testing and plays a role in instrument manufacture and preflight calibration as well. See Pinch, "Testing, One, Two, Three, Testing"; Downer "When the Chick Hits the Fan."

19. Field notes, March 2, 2006. This was a particularly generative session, in which I observed a training interaction between two other calibrators.

20. Galison, "Judgment against Objectivity."

21. Shapin, "Invisible Technician."

22. On scripts see Akrich, "De-scription of Technological Objects." Whereas the actor-network theory notion of a script draws attention to the variety of actors, human and nonhuman, that need to be in place for a technological script to work, my focus here is on the prescribed and routinized way these students interact with the sociotechnical system.

23. *Pancam Calibration Procedures for Extended Mission*, Version 1.29. This document is evolving and maintains remnants of earlier versions, sometimes including instructions that were more important in the early days of the program (e.g., what to do with high-priority image data for JPL while new images are coming in) that are now ignored, or missing steps (e.g., "6.02 [Step removed]").

24. Field notes, March 2, 2006.

25. Field notes, March 5, 2006.

26. Field notes, March 2, 2006.

27. On black boxes, see Latour, *Science in Action*. The software was written in-house by a graduate student on the mission who has since left the lab but who also wrote a program that can display all calibration software scripts on request. In practice, however, this program is very rarely consulted, since many PCC members either do not know enough IDL to read the scripts or were the authors of the scripts themselves.

28. For example, Barley, "Technology as an Occasion for Structuring"; Barker and Downing, "Word Processing and the Transformation of Patriarchal Relations of Control in the Office."

29. Orr, *Talking about Machines*.

30. Doing, "'Lab Hands' and the 'Scarlet O.'"

31. Notably, there is upward mobility in calibration: experience as a calibrator opens the door to ground-floor mission participation for a wide range of students. Several calibrators I met went on to graduate training in the field of planetary science, inspired by their experience on the PCC. Perhaps because of this desire to use the PCC as a learning experience, on reading an earlier draft of this chapter the PCC leaders decided to host a "miniworkshop" "to address the reasons for the various steps in the PCC pipeline, changes to the PCC procedures since the start of the mission, and questions or concerns that you have about calibration." The e-mail invitation to this workshop emulated the rules of engagement on the mission at large, repeating frequently that all questions were welcome so that all could hear the answers (PCC internal e-mail, January 29, 2008).

32. Pancam Calibration Procedures, 7.01.

33. Lynch, "Discipline and the Material Form of Images."

34. These procedures are described in detail in Bell et al., "In-Flight Calibration."

35. Pronounced "I over F." One of my instructors explained this as the "ratio of radiance seen versus perfect radiance" (Field notes, March 5, 2006). However, I found much variation among Rover team members about what the acronym stands for.

36. Described in chapter 3.

37. Constructing dust and even sunlight as artifacts recalls Lynch's ethnomethodological study of the construction of artifacts through work and talk in a biology laboratory (*Art and Artifact in Laboratory Science*). Of additional interest are discussions of programmed corrections to observations in Lynch and Edgerton ("Aesthetics and Digital Image Processing") and

Lynch ("Laboratory Space and the Technological Complex"). I am thankful to the Pancam payload element lead and his postdoctoral student for their explanations and demonstrations of flatfielding.

38. Field notes, March 2, 2006.

39. Unlike "normal accidents" (Perrow, *Normal Accidents*), "nominal anomalies" are known and acknowledged blind spots that members believe reflect expert knowledge of the peculiarities of their instrument rather than catastrophic events with risky technologies that arise from organizational norms and "just following orders."

40. This is a central theme in recent science studies literature, expressed by authors such as Daston and Galison (*Objectivity*), who describe how scientists are disciplined by their procedures to guarantee objectivity, or by Donna Haraway (*Simians, Cyborgs, and Women*; *When Species Meet*), who draws our attention to the mutual entanglement of objects and subjects and therefore pushes for an analytical language that does not rely on dichotomous categories to begin with. This book contributes to such scholarship by emphasizing that even as members of the Rover team produce images that draw Mars in particular ways, at the same time they produce the local social order of their team. Images of objects (Mars) are concomitantly images of subjects (the team of rovers and Rover scientists) too.

41. Calibration is not "doing science." A moment in my fieldwork made this distinction and its status clear to me. When a picture I had calibrated appeared on the cover of *Science* magazine (vol. 313, no. 1792, September 8, 2006), I excitedly exclaimed, "I calibrated this image!" The Rover scientist in the room with me gently but firmly replied, "And I processed it." The implication was that calibration is distinct from the work of scientific analysis, the work of image processing, to which both intellectual and artistic credit were due.

42. Daston and Galison, "Image of Objectivity"; Galison, "Judgment against Objectivity"; Daston and Galison, *Objectivity*.

43. The reference here is to Donna Haraway's critique of the "view from nowhere" or the "God's-eye view" (*Modest_Witness*). However, as I will argue later in this book, the rover's-eye view makes for a very situated perspective on the Martian landscape.

44. Latour and Woolgar, *Laboratory Life*, 76. This account adds nuance to the story of the transparency of inscriptions, whereas we might otherwise have claimed, with Latour, that images efface the process of their production. This is especially evident in the contrast between the epistemic status of the raw images versus the status of the calibrated ones. NASA posts all the raw image data on its Rover website, but the calibrated versions are released several months later. As my instructor explained, if anyone tried to conduct scientific photometry or spectral analysis with the raw images, the results would be flawed—even though these images constitute firsthand witness reports, the closest thing to a trustworthy inscription produced by an inscription device on the mission. So inscriptions from the camera must be consistently monitored and modified or else they cannot be taken as the object speaking for itself. It is certainly true that actors' accounts of the Pancam include the direct self-registration of natural effects—photons onto a CCD plate—as discussed in chapter 3. But accounts and practices rarely align: in fact, practices like calibration are necessary to support these accounts. Only when images are calibrated—*drawn as* trustworthy observations—can mission scientists *see* this data *as* evidence, uninhibited by instrumental artifacts or observer bias.

45. As discussed in the chapter 8, all raw images are immediately released to the public and can be viewed at http://marsrovers.jpl.nasa.gov/gallery/all; calibrated images are released to NASA's Planetary Data System in three-month packets. Caltarget images are available for

download, but the scripts that enable their interpretation and calibration are not released publicly, prompting much discussion among amateur sites as to how to calibrate the newest images themselves. This may seem counterintuitive at first: surely these images are too raw for public consumption and present a vulnerable side of the Rover program. However, the policy accomplishes the dual goals of upholding the scientific norm of communalism while still restricting access to the calibrated images to a core set, bounded by the Rover team, who "certify new knowledge" (Collins, *Changing Order*, 143) about Mars. Still, a three-month turnaround for public release of calibrated images is considered extremely fast: many high-profile space missions guard their data closely until their team has amassed enough publications. In such an environment, the early decision to release the Rover team's images as soon as they hit the ground was considered a rare gesture and has influenced other missions since then. On this topic, see Vertesi and Dourish, "Value of Data."

Chapter Three

1. The following quotations and descriptions are taken from my ethnographic observations on June 12, 2007.

2. Such techniques are recognized by scholars in science and technology studies as central to image analysis and interaction, as in Morana Alač's discussion of gesture, talk, gaze, and cursor sweeps and Catelijne Coopmans's description of mammogram analysis software salesmen deploying artful revelation of different aspects of the data. Alač, *Handling Digital Brains*; Coopmans, "'Face Value.'"

3. Coulter and Parsons, "Praxiology of Perception."

4. This point has been with us in the history, philosophy, and sociology of science for a long time. Ludwig Fleck calls scientific images ideograms: "graphic representations of certain ideas and certain meanings . . . where the meaning is represented as a property of the object illustrated" (Fleck, *Genesis and Development of a Scientific Fact*). Donna Haraway's concept of "figuring" also establishes a relation between how we materially/discursively draw objects into boundaries, relationships, and dualities consistent with what and how we know about those objects (Haraway, *Modest_Witness*). Like these scholars, my concern here is the production and reproduction of such ideograms and figurations, with their inscribed ways of seeing and analyzability, particularly how they link scientific ways of seeing and knowing with representational work.

5. This is a common approach in science and technology studies, where ethnographers emphasize the importance of the stories scientists tell about the instruments they use. In cases such as neuroscientists describing their PET scanners, the analyst can learn about the particular kinds of value practitioners place on their image data and what they can and cannot show: see Dumit, *Picturing Personhood*; Beaulieu, "Images Are Not the (Only) Truth." The present discussion is based not only on extensive conversations and observations with Rover team participants, but also on participation in an undergraduate planetary image-processing class taught by the Pancam Payload Element Lead in fall 2005. I am grateful to the instructor for allowing me to attend his class, Astronomy 310, as part of my fieldwork on the mission.

6. A widely used textbook in planetary image processing presents the water bucket analogy as a story that unifies image processors around the CCD and a vocabulary about the direct relation between photons and pixels: "Buckets represent pixels on the CCD array, and a rainstorm provides the incoming photons (rain drops). Imagine a field covered with buckets aligned neatly in rows and columns throughout the entirety of the area. . . . After the rainstorm (CCD integra-

tion), each bucket is transferred in turn and metered to determine the amount of water collected. A written record (final CCD image) of the amount of water in each bucket will thus provide a two-dimensional record of the rainfall within the field." Note that the water bucket analogy leaves no room for ambiguity. The point of the story is not that a drop of water in a bucket then mixes with and becomes indistinguishable from others in a mass of water, but that the unit of the rain-drop is both precisely recorded and maintained. Unlike the case in brain scanning, then, pixels do not represent statistical values. To planetary image processors, the pixel ("picture element") is a precise, direct quantification of the light that hits the CCD detector (Howell, *Handbook of CCD Astronomy*, 8).

7. Rover scientists who work with these images frequently view pixel data in both a numeri-cal and a pictorial way. As one graduate student I interviewed put it, pictorial images of Mars are "really just a visualization tool: all you get from the CCD is a bunch of numbers" (Interview, Thomas, September 12, 2006). It is always possible to view the raw pixel values listed as a stream of numbers in a text document. However, because Pancam images contain 1,024 rows and 1,024 columns of pixels, the sheer volume of numbers can quickly become overwhelming. Many scien-tists prefer instead to plot pixel values on a graph or to apply a mathematical function to them, as I will describe in chapter 7. There is also little anxiety about how much to believe the numbers versus the pictures. Scientists frequently describe and deploy both views as simply different ways of seeing the same data. Compare this way of working with digital images with those described in Beaulieu, "Images Are Not the (Only) Truth"; Lynch, "Science in the Age of Mechanical Repro-duction"; Joyce, "From Numbers to Pictures."

8. See, for instance, Kinch et al., "Dust Deposition."

9. Wittgenstein, *Philosophical Investigations*, 195.

10. To produce these pictures, scientists like Ben work with a suite of tools in their image-processing software of choice, ranging from hand programming in an image-processing language like IDL, to the Pancam software suite, to the USGS's software ISIS, to commercially available tools like ARC-GIS or even Adobe Photoshop. While each is slightly different in its focus, all these programs allow scientists to select several frames they wish to combine, dictate which color channels to assign to which frames, and tweak the resulting color image. Once the program is running, the scientist then aligns the image taken through a red filter with the red channel in the processing software, the bluest filtered image through the blue channel, and the greenest image through the green channel.

11. Images generated in some form of true color have a variety of names across the planetary sciences, distinguishing different algorithms that encode decisions about what that human sensitivity is. On the Rover mission, true-color images released to the public are combined ac-cording to an algorithm developed at Cornell called Approximate True Color (ATC), so named to emphasize the very constructed code manufactured by a human decision made on Earth instead of from direct experience of Mars. On other missions this algorithm may be differently construed and differently named, as in, for example, the "Natural Color" images released by the Mars Reconnaissance Orbiter's HiRISE camera. I describe uses of ATC in chapter 8.

12. Interview, Martin, June 8, 2007; emphasis in the original.

13. Interview, Kwame, February 16, 2007.

14. Wittgenstein, *Philosophical Investigations*, 167.

15. Wittgenstein, *Philosophical Investigations*, 171. Quotation marks and triangle in original.

16. Laboratory observation, December 4, 2006.

17. Laboratory observation, December 4, 2006.

18. On intentionality in image making, see Baxandall, *Patterns of Intention*. In this case the purposeful selection of filters that permit only particular kinds of views makes identifying scientists' intentions less of a dark art than might at first be imagined.

19. Interview, Julie, June 12, 2007. Ben resists the label of "pixel pushing" as "somewhat of a derogatory term . . . applied to the drone-like process of running canned computer routines to generate standard images." Personal correspondence, June 22, 2009.

20. Interview, Ben, June 11, 2007. This statement resonates strongly with the postphenomenological work of Don Idhe (*Postphenomenology and Technoscience*), a philosophy that explores the relationships humans develop with our instruments that extend and literally incorporate instrumental ways of seeing and knowing. It also points to how work with the Pancam instrument is one of the first and most instrumental ways of learning to "see like a Rover." Importantly, this postphenomenological sensorium also includes human interactions with software. I will return to this theme in chapter 6.

21. Laboratory observation, December 4, 2006.

22. The following descriptions and quotations are taken from my ethnographic interview and observations with Susan, June 18, 2007.

23. Team meeting, February 14, 2007.

24. I will discuss the importance of laboratory work to corroborate image work in chapter 7.

25. As Jennifer Tucker relates, Lowell and his research group had been struggling with these issues of photographic visibility since the early part of the decade. Lowell's invitation to Berlin came on the heels of his invitation to exhibit at the Royal Society in London in 1905 and his medal from the Royal Photographic Society in 1907, but despite these successes, the photographs' graininess was still considered a sticking point for the credibility of the canals. See Tucker, *Nature Exposed*, esp. 215-33.

26. V. M. Slipher to C. O. Lampland, January 30, 1909, Lowell Observatory Archives, Flagstaff, Arizona. For Lowell's opinions about the habitability of Mars, see Lowell, *Mars*.

27. There is also evidence that Lowell experimented with the technique of photographic composites, combining photographs of Mars with photographs of his drawings of Mars in order to help observers see what he could see. I am grateful to Antoinette Beiser at the Lowell Observatory Archives for access to this photographic and correspondence material.

28. The literature on Galileo is vast. On this example in particular see Edgerton, "Galileo"; Winkler and Van Helden, "Representing the Heavens"; Biagioli, *Galileo's Instruments of Credit*, 105-11; Bredekamp, *Galilei der Künstler*. I thank Eileen Reeves for her generous assistance with navigating related materials.

29. Ptolemaic astronomy, prevalent at the time, stated that the heavens were made of a different element than Earth. Called the fifth element, or quintessence, this element was unlike the earthly elements of earth, air, fire, and water in that it was perfect and moved in perfect circles. According to many of Galileo's contemporaries, the spots on the moon were not pockmarked imperfections but symbols of the moon's perfection, like colored patches on fine marble. Those astronomers who were sympathetic to Copernicus, however, saw things differently. If Earth was a planet like any other, there was no reason to assume that physics on Earth was any different from the physics on other worlds. The literature on the Copernican and Ptolemaic systems is also vast, but I refer interested readers to Kuhn, *Copernican Revolution*.

30. I make passing reference to these cases here not because I believe their histories are cursory. My own work in the history of seventeenth-century astronomy is concerned with how very complex and entangled these cases can be (cf. Vertesi, "Picturing the Moon"). Rather, I offer

these as analogs to demonstrate the analytical purchase of the *drawing as* framework as it can be transported and used in other domains aside from the digital. Reference to such cases can also illuminate the importance of the work on the Rover mission as operating in a broader context of astronomical understandings, despite my ethnomethodological focus on local, bounded cases.

31. On Harriot's maps see Edgerton, "Galileo"; Bloom, "Borrowed Perceptions"; Pumfrey, "Harriot's Maps of the Moon." On Galileo's influence on other forms of lunar representation, see Kemp, "Maculate Moons" (in his *Spatial Visions*, 40-41). For a later example of competing lunar representations informed by different theories about the moon, see Vertesi, "Picturing the Moon." On the mobility of images, see Latour, "Visualization and Cognition."

32. Interview, Sam, May 22, 2007.

33. End of Sol meeting, September 6, 2006.

34. Team Meeting, February 14, 2007.

35. Interview, Tom, June 26, 2007.

36. Making something "pop out" while excluding or silencing other perspectives and scientists' talk of "throwing something out in order to see" recall the familiar Foucauldian theme of discipline in scientific representation (Lynch, "Discipline and the Material Form of Images"). Even the language used to discuss image processing betrays a kind of violence as pixels are "pushed" or "stretched" into conformity; another scientist talked about the need to "pull information out of digital data"; and as Ben explained to me in our interview, "you need to pound the data to this level to be able to see the secondary differences between things." But as I described in the example of calibration, being a member of a scientific discipline not only requires disciplining unruly objects into compliance. At the same time, one is oneself disciplined into being a particular kind of scientist. Appealing to these disciplinary distinctions—whether geology, geochemistry, or engineering—also produces those very distinctions in the context of interaction.

37. Here I refer to science studies work on multiple ontologies, exemplified in Annemarie Mol's influential book *The Body Multiple*. As Mol traces encounters with atherosclerosis across a hospital setting, her ethnographic account explains how the disease is differently enacted in different parts of the hospital according to different tools, techniques, and practitioner identities. She then explains how these multiple atheroscleroses are made singular through actors' work of coordination and translation. Along with other scholars, Mol recommends that we conceptualize scientific objects as being multiple so that we can expose how scientists' practices stitch together distinct object ontologies to associate them as singular entities. But just as Mol observes organizational practices at her hospital site that produce and coordinate a disease made multiple, the organizational practices I observe on the Rover mission keep Mars very much in the singular. The visual work done by members of this mission team is part and parcel of their unified sociotechnical stance and is both reflected in and projected from their image production. As I described in chapter 1, coordination work is not a post hoc construction of an intrinsically fragmented entity or field; instead it is built into the image's very acquisition. It is likely, then, that what makes objects multiple or singular is not (only) a function of the representational modes or disciplinary practices, but is instead enacted through the local social order and associated organizational practices that produce such coordination.

38. Wang and Ling, "Ferric Sulfates on Mars"; Wang et al., "Light-Toned Salty Soils"; Squyres et al., "Detection of Silica-Rich Deposits on Mars."

39. A growing body of scholarship in science and technology studies, largely drawing from anthropology, actor-network theory, and feminist philosophy of science, deploys the ontologies literature and vocabulary in analysis (cf. Haraway, *When Species Meet*; Suchman,

"Subject Objects"; Barad, *Meeting the Universe Halfway*). To readers steeped in this literature, my references to "drawing analytical objects as natural objects" may seem to maintain an a priori distinction between objects and subjects or to confuse my participants' categories with my own analytical ones. This is not my aim. Instead, I want to point to how the practical work of visual construal (re)presents objects of scientific observation as already interpreted, thereby producing subject-objects and emic ontologies through the activity of depiction. I therefore call attention to how representational practices do not present "the things themselves" but instead present actors' interpretative work in a way that is seamlessly integrated with the world they depict. Certainly this underscores the well-known point that any analysis of scientific images, whether contemporary or historical, cannot analyze images or their representational qualities by comparing them with a "world out there" or with objects as we know and make them today. But it also points to an intermingling of epistemological and ontological work in representational practice. Thus my approach to ontology is one that restates the central concern of this book: to show how practices of making the world are bound up in socially ordered modes of interaction, including those interactions related to techniques of visual representation.

Chapter Four

1. End of Sol meeting, March 28, 2007.

2. Field notes, September 2006. This moment is reminiscent of a poignant scene in Mol's ethnography when a vascular surgeon notes that angiograms are "like a road map . . . [b]ut we treat patients, not pictures" (*Body Multiple*, 93). Important for understanding images and interactions, image processing and the interpretative work of *drawing as* informs interactions yet is distinct from them. My emphasis here is on the importance of those mapping techniques and practices that lay out and inform possibilities for interaction.

3. For example, Cosgrove, *Social Formation and Symbolic Landscape*; Cosgrove, *Geography and Vision*; Wood, *Power of Maps*; Harley, *New Nature of Maps*; Scott, *Seeing like a State*.

4. Wood, *Power of Maps*.

5. The notion that "maps are propositions" is put forward by Krygier and Wood ("Ce n'est pas le monde").

6. Hacking, *Representing and Intervening*.

7. I have elsewhere used an empirical study of the London Underground map to explore similar principles. Residents of London regularly use the city's famously nongeographical subway map not only to plot their course in the transit system, but also for understanding and interacting with their city in general. "The question thus changes from whether or how well the visualization represents the object, to how the visualization constructs the object for interaction" (Vertesi, "Mind the Gap," 11).

8. On drawing observations together, see Latour, "Drawing Things Together"; Latour, "The 'Pedofil' of Boa Vista."

9. Team Meeting, July 7, 2007.

10. A description of geological maps and their history at the end of the eighteenth and early nineteenth centuries is in Rudwick, "Emergence of a Visual Language," esp. 159–64. Human factors researchers who studied the Rover team during the primary mission period saw Athena scientists like Joseph printing out images and drawing on them; they provided the scientists with a long table where they could lay printouts of recently acquired images and gather around them to discuss their interpretations.

11. Although geological mapping has been extended from Earth to other planets in the solar system, it is generally recommended that the eye of the planetary mapper be trained with terrestrial experiences, since this is how young planetary scientists can best acquire expertise about how features on the ground can be recognized from space. This training involves learning to read orbital images, drawing on them to transform them into maps that identify particular types of terrain, stratigraphic layers, or mineral deposits, and taking these orbital images into the field, walking carefully around the area on Earth to better understand how what is on the ground is seen from space. As Don Wilhelms explains in Greeley and Batson's classic textbook, *Planetary Mapping*, translating Earth-based experiences to other planets is complicated by the fact that while maps of Earth moved from ground-based surveys to the synoptic view provided by aircraft and satellites, planetary images have the opposite approach, moving from fly-by images, to images from orbit, to ground-based rovers. I will return to many of these issues in chapter 7. I am grateful for conversations with Larry Crumpler, Jeff Moore, Ken Tanaka, and Don Wilhelms on the topic of geological maps.

12. Team Meeting, July 7, 2007.

13. Team Meeting, July 7, 2007.

14. The notion of "drawing things together" owes much to Bruno Latour. To move inscriptions along a chain of translation from direct observation to graphic traces, investigators assemble a "chain of inscriptions" to draw together a variety of local observations into a graphical whole. In the Rover case, such investigators do not return home from afar or send samples to a "center of calculation." Rather, the process of assembly is continuous, digitally mediated, and developed iteratively for, with, and through further rover observations (Latour, "Drawing Things Together" and "The 'Pedofil' of Boa Vista").

15. Interview, Sam, May 24, 2007.

16. End of Sol meeting, September 6, 2006.

17. *Opportunity* SOWG meeting, sols 1005–6, November 20, 2006.

18. This move is reminiscent of "professional vision" (Goodwin, "Professional Vision"); "selection" (Lynch, "Discipline and the Material Form of Images"); and "fixating visual evidence" (Amman and Knorr-Cetina, "Fixation of [Visual] Evidence").

19. *Opportunity* SOWG meeting, sols 1029–31, December 14, 2006.

20. Knorr-Cetina and Amman, "Image Dissection," 280.

21. End of Sol meeting, April 4, 2007.

22. *Opportunity* SOWG meeting, sol 1175, May 14, 2007.

23. End of Sol meeting, September 6, 2006.

24. MOC was the highest resolution available at the time, and images from its catalog were used for strategic planning. During my fieldwork, however, the Mars Reconnaissance Orbiter (MRO) camera came online, featuring not only high-resolution cameras that could spot the rovers from orbit, but also team members shared between the MRO team and the Rover team. Orbital images were incorporated with increasing frequency into both strategic and tactical planning sessions, since team members could now request high-resolution orbital images of specific locations from MRO to assist their local planning and interpretation for *Spirit* and *Opportunity*.

25. End of Sol meeting, September 6, 2006.

26. End of Sol meeting, September 6, 2006.

27. End of Sol meeting, September 6, 2006.

28. Red and blue dots placed on screenshots are regularly used in both planning and recording MiniTES operations. A scientist will send the MiniTES PUL a screenshot image with

labeled red dots on a Navcam or Pancam image to show where to place the observation; within their software, the MiniTES PUL will closely approximate the location of the screenshot dots to provide the pointings for the MiniTES on Mars; then a MiniTES worker will create a screenshot JPG file of that operations software image (featuring blue dots) to record the location of the targets for the sake of recording them in the Planetary Data System. I am grateful to interviews with several MiniTES PULs and a site visit to their operations center in Arizona for their descriptions of this work.

29. An analysis of rover site names in this vein is unfortunately beyond the scope of this book. On the cartography of planetary bodies, see Lane, "Geographers of Mars"; Vertesi, "Picturing the Moon."

30. Reportedly, the team once used the SOWG meeting room to watch the World Series.

31. I am grateful for an interview with Roxana Wales (February 20, 2007) about the early history of naming on the team. For more on targeting and the use of nomenclature to distinguish targets, see Wales, Bass, and Shalin, "Requesting Distant Robotic Action." I am also grateful for a discussion with Mark Powell and Jeff Norris (February 24, 2007) about targeting as a question of visual information processing. See also Powell et al., "Targeting and Localization."

32. Most of the time these two principles of naming are combined. Names stand as a record of both team achievement on Earth and robotic achievement on Mars. Occasionally, however, discrepant names indicate a divergence between team activity and robotic activity, as in the case of Innocent Bystander, described below.

33. Names also work as a classifying system that ties discrete observations together across Martian space and time under the rubric of a geological hypothesis. For more on classification, ontology, and coordination, see Bowker and Star, *Sorting Things Out*; Mol, *Body Multiple*.

34. *Spirit* SOWG meeting, sols 1234–36, June 22, 2007.

35. *Opportunity* SOWG meeting, sols 1128–29, March 27, 2007.

36. Interview, Tom, June 26, 2007.

37. DEM data is sometimes represented as an undulating landscape composed of squares, much like graph paper. However, the very Pancam images that scientists used to determine the topography can also be "draped over" this graphic space to give an immersive view of the rover's environment. One engineer, Jesse, described this as using the original image as "skin on the texture map" (Interview, Jesse, February 15, 2007).

38. Interview, Ying, June 26, 2007.

39. Interview, Yao, June 26, 2007.

40. In yet another aspect, DEM data can also be *drawn as* an even more spectacular vision of Mars; draped with Approximate True Color Pancam or orbital images, image processors use DEM to create dramatic "fly-through" movies that simulate soaring above the rover landing sites or through Valles Marineris canyon. Although I saw scientists like James accomplish this work, it is also performed by engineers at JPL in facilities similar to those of a Hollywood studio, where the processing lab equipment, techniques, and even personnel overlap with those of the local film industry.

41. Interview, Li, June 27, 2007.

42. Interview, Bo, June 27, 2007. This points again to the situated nature of these maps not as large-scale cartographic productions, but as local mapping specific to immediate interaction.

43. Interview, Jesse, February 15, 2007.

44. Hacking, *Representing and Intervening*.

Chapter Five

1. End of Sol meeting, September 19, 2007.

2. On these and other practical aspects of map production in science and statecraft, see Scott, *Seeing like a State*; Lane, *Geographies of Mars*; Vertesi, "Picturing the Moon"; Burnett, *Masters of All They Surveyed*; Winichukl, *Siam Mapped*.

3. Notably, the Mars rovers are a product of NASA, a state agency with considerable authority. But Rover mapping as I observed it did not travel beyond the mission. Certainly, the practices I detail here are the practices of producing a collectively assented view of Mars from among members of a state-selected group of fortunate individuals. But the limitations of the rovers' mobility make them ill suited to state mapping projects, which are more often undertaken and published by central state actors such as the US Geological Survey, even in the case of other planetary bodies. Examining these backstage practices thus exposes the kinds of interactions essential to knowledge production that occur in collectively organized environments, whether supported by a state agency or located elsewhere.

4. See Polletta, *Freedom Is an Endless Meeting*. Disagreements on the Rover mission tend to be contained, with limited emotional engagement, and highly depersonalized. Raging conflict or personal disagreement is extremely rare. In none of the cases I describe below did people raise their voices, stop listening, storm out of the room, disengage, or make something personal. Even cases in which two groups saw things quite differently, such as scientists and engineers, were relatively rare and were cause for concern among the team members.

5. During the distributed operations phase of the mission, these conversations occurred virtually. I happened to witness an informal one in person, pictured in figure 5.2.

6. End of Sol meeting, March 7, 2007.

7. End of Sol meeting, January 31, 2007.

8. End of Sol meeting, January 31, 2007.

9. End of Sol meeting, September 12, 2007.

10. End of Sol meeting, January 24, 2007.

11. Maps like Roger's are routinely displayed in the LTP reports at the outset of the daily SOWG meetings, serving quite literally as "the big picture" to anchor the SOWG's tactical conversation within the strategic context. They also remind team members of what they have already agreed to do, establishing their authority through the collectivity of the underlying discussion and serving as the mission's recorded memory. In fact, so standard is this routine of placing these images into the LTP reports to remind the team of their evolving location, goals, and decisions that a fake image was once traced in LTP reports for up to a week. This was a Photoshopped joke image of a four-digit odometer reading "9999" nestled among the rover's solar panels, commemorating *Opportunity*'s impending ten kilometer driving mark. This image continued to be included in subsequent LTP reports not because team members were not familiar enough with the rover's instruments to recognize the joke, but rather because it was such a powerful reminder of the team's achievement to date that it deserved inclusion alongside other annotated reminders of collective achievement. On the analysis of scientific humor, see Mulkay and Gilbert, "Joking Apart."

12. *Spirit* SOWG meeting, sols 1100-1101, February 5, 2007.

13. *Spirit* SOWG meeting, sols 1100-1101 February 5, 2007.

14. Note that the two-to-one situation here is resolved not through voting but through Jane's agreeing to accept the alternative. The Chair did not declare the decision made until she did so.

In my years with the mission I never saw voting performed on the team. In fact, the team actively discourages voting as a way of solving disagreements, arguing that it silences minority perspectives and therefore conflicts with the team's value of listening. In conversation, more than one team member recalled a moment early on in the mission when one rover's SOWG team adopted voting as a strategy; when the PI joined in that rover subgroup one day, he was said to have exclaimed, "Voting? We don't vote here!"

15. End of Sol meeting, September 6, 2006.

16. End of Sol meeting, September 6, 2006.

17. End of Sol meeting, November 29, 2006.

18. End of Sol meeting, November 29, 2006.

19. Note in this example that the annotations in the image can stand only if they are not challenged by other members of the team. Discussing this incident with another team member a few years later, this team member explained William's interjection as essential and even expected according to team norms: "If you label the slide, it just snowballs. You gotta cut someone off [if you disagree]. This is Stewart: people trust what he says. Then LTP Leads reuse the slide, and it becomes part of the lexicon. It takes so much effort to correct [a mistake] down the line. If it works its way into papers, the consequences get progressively more dire" (personal conversation, August 25, 2011). As this scientist explained it, it was responsible to maintain some interpretative flexibility in the image, in case the issue appeared conclusively resolved prematurely. Premature interpretations would translate "down the line" into a standardized view and accepted fact. Images that do make it "down the line," with their visual interpretations drawn onto them, must represent points of agreement among the team.

20. An instructive comparison here is Charles Goodwin's study of a group of oceanographers who must work together and coordinate their visual and laboratory practices to make decisions about where their ship should go and what data it can collect (Goodwin, "Seeing in Depth"). He describes how the different disciplinary conventions associated with each group on board—oceanographic physicists, geochemists, and even sailors—affect how each scientist perceives and interprets the visual field. These different actors use image work to coordinate and organize their activities and ultimately produce knowledge of the Amazon River basin. Here I want to bring our attention to the process of coordination as part and parcel of the team's local social order.

21. Knorr-Cetina, *Epistemic Cultures*.

22. Note that this isn't the same dust as removed by calibration. Calibration removes dust effects in the atmosphere, but the dust in question here is coating the prospective targets.

23. I am grateful to Rover team members for their recollections of the landing site selection process; also to project scientist John Grotzinger for permission to attend the Mars Science Laboratory landing site workshop in September 2008 to get a feel for the process and pressures of selecting a site. Many of the documents related to the rovers' landing site selection are available at http://marsoweb.nas.nasa.gov/landingsites/index.html# (accessed November 17, 2008).

24. Once the rovers arrived on the ground, however, the two sites reversed their appeal: Meridiani's craters invited geomorphological analysis, while Gusev's volcanic plain gave way to the mineralogy of Home Plate.

25. *Spirit* SOWG meeting; date withheld.

26. The quotations and description below are from the End of Sol meeting on October 30, 2007.

27. Rogan is a target name for a rock representing a geological stratum of interest at Home Plate. Consistent with other targets in the area, it was named after Charles Wilbur "Bullet" Rogan, a player for the Kansas City Monarchs, to celebrate the United States' Black History Month.

28. This End of Sol meeting revealed a divide between scientists and engineers on the mission: a divide that, as I explained in chapter 1, the team works hard to bridge. As in any boundary negotiations, however, the members must constantly draw attention to the distinction in order to surpass or negate it. For example, the team draws a sharp boundary between "science" and "operations" in the context of work roles and responsibilities, yet it is this same distinction that enables something like the Science Operations Working Group (SOWG) meeting—the combined meeting of the two domains—to be so powerful in the context of the organization. Yet the two sides also have different expertise, and both draw and see Mars quite differently to reflect those foci. Local knowledge-management practices require that the multiple sets of images and views of Mars be placed alongside each other to inform subsequent interactions, yet those are not always considered equal in the context of science or of operations decision making. For example, when one scientist attempted to posit a possible drive path from an image, a Rover Planner interjected, "I think it might be wisest to leave the science to the scientists and leave the engineering to the Rover Planners." When the scientist protested, the Chair shut the conversation down politely but abruptly with a "No, [scientist], please" (SOWG meeting, date withheld). While there is tremendous exchange between the two sides, their construal as separate sides of the same coin enables that exchange.

29. Lynch, "Externalized Retina."

Chapter Six

1. Liz, observation, February 1, 2008.

2. Liz, observation, February 1, 2008.

3. Alač, "Working with Brain Scans"; Myers, "Molecular Embodiments"; Prentice, *Bodies of Information*; Idhe, *Postphenomenology and Technoscience*; Rosenberger, "Perceiving Other Planets."

4. I remain agnostic about how or whether the rovers "really see" the planet. What was observable to me, and analytically interesting, is the formulation of a visual convention that develops a bodily, intersubjective experience of *seeing like a Rover* and which is shared by the team as a marker of membership. Throughout this chapter, then, I do not address the rover's own point of view, but rather explore a crafted and constructed visual and embodied experience that is made, circulated, and part of sense-making on Earth.

5. Astronomy 310 field notes, October 2005.

6. Interview, Mark, February 15, 2007. During my work on the team, the rovers received several upgrades to this onboard visual analysis system, enabling them to drive farther without intermediate imaging.

7. Interview, Mark, February 15, 2007.

8. Goodwin, "Professional Vision."

9. Note the similarity to Sarah's image work and dismissal of South Promontory as a winter haven for *Spirit*, despite the scientific reasons to travel there. Sarah used stereo anaglyphs to show that "although it looks flat" the drive to the outcrop was "full of fairly large rocks." She also used stereo Pancam data to produce a lily pad map to find a drive path, and DEM data to determine

slopes. Seeing like a scientist, as Joseph did, South Promontory was a compelling place to explore; but seeing like a Rover, it was impassable.

10. On robotic anthropomorphism, see DiSalvo et al., "All Robots Are Not Created Equal"; DiSalvo and Gemperle, "From Seduction to Fulfillment." With *technomorphism* I do not mean to counter the rich literature in science and technology studies on the co-constitution of subjects and objects, or the blending of the cyborg body; nor do I wish to invoke technological determinism. However, during object co-construction, it is important to pay attention to exactly which resources (talk, gesture, or visualization) actors deploy to draw and efface distinctions in the management of shifting robotic and team identities (see especially Alač, "Moving Android"; Suchman, "Subject Objects"). Technomorphism articulates one of those resources: how my ethnographic participants narrated and performed their experiences on Mars through that of the robotic body, not always the other way around.

11. Interview, RAT operator, August 13, 2007.

12. Interview, Mark, February 15, 2007.

13. Interview, Jordan, February 15, 2007.

14. Jude, personal communication, September 2006.

15. Interview, Mark, February 15, 2007.

16. *Opportunity* SOWG, sol 927, September 1, 2006.

17. Or communicating from an expert to a novice, as in Alač's semiotic analysis of gesture and visibility in the case of brain scan analysis ("Working with Brain Scans").

18. End of Sol meeting, May 16, 2007.

19. Interview, MiniTES PUL, June 6, 2007.

20. Jude, personal conversation, field notes September 2006.

21. Interview, May 25, 2007. On the rovers' gender, see 187–90.

22. Interview, February 15, 2007.

23. Personal conversation, July 2007.

24. My thanks to Leigh Star for suggesting that I include a description of my own embodied experience in my analytical narrative.

25. Myers, "Molecular Embodiments," 62.

26. Alač, "Working with Brain Scans."

27. Prentice, *Bodies of Information.*

28. Merleau-Ponty, *Phenomenology of Perception.*

29. See Clancey, "Becoming a Rover"; Schairer, "Diffused Embodiment, Extended Visions."

30. Liz, personal conversation, February 6, 2008.

31. Interview, Roger, May 24, 2007. I can report a similar experience in my ethnographic work. When I visited the Grand Canyon with a Rover team member stationed at the USGS, the geology did not come alive for me until it was described in rovers' terms (e.g., This dust is like the dust on *Spirit*'s solar panels; this is an example of cross-bedding like that at Cape Verde). Visiting Meteor Crater in Arizona, I was disappointed in the view compared with Victoria Crater— until I crouched down to rover height (field notes, June 2007).

32. This resonates with the postphenomenological work of Don Idhe and his colleagues, who suggest that not only do we shape our instruments, but they also shape us in our forms of perception. Idhe, *Postphenomenology and Technoscience.*

33. Merleau-Ponty, *Phenomenology of Perception,* 67–68.

34. Hans Radder, *World Observed/The World Conceived,* 70, 73–75.

35. This particular entanglement of humans and machines resonates with science studies theorist Donna Haraway's notions of the cyborg or of interspecies relations. In Haraway's account, the dichotomies of human/machine or nature/culture are disrupted by the intimate, embodied, and boundary-crossing relationships offered by cyborgs or companion species. Moments where Rover team members talk of learning to see like their rovers recall Haraway's description of learning to see the world with and as her dog through obstacle course training: both inspire rearticulation of the binaries and boundaries between human and nonhuman. Lucy Suchman brings Haraway's notion to bear on her concept of configuration in human-machine relationships. In her studies of work with photocopiers or with artificially intelligent robots, Suchman shows how humans and machines are mutually articulated through their interactions (Haraway, *Simians, Cyborgs and Women,* and *When Species Meet*; cf. Suchman, *Human-Machine Reconfigurations*). The embodied representational modes, talk, action, and narrative that I describe in this chapter draw similarly intimate, awkward, and boundary-crossing connections between Rover team members and their robots. At extreme distances, these are played out as interaction with and through images. While it is not my method here, actor-network theory might also suggest an approach to the symmetry and mutual articulation of humans and machines, where humans and machines are entangled equally as actors and resources in a network that produces knowledge. My ethnomethodologically inspired focus remains on how team members themselves account for these intimate bodily relations, and how these accountings are intertwined in their visual and ordered practices.

36. Knorr-Cetina and Bruegger, "Market as an Object of Attachment."

37. Liz, personal correspondence, May 27, 2008.

38. *Spirit* wake, July 18, 2011.

39. The notion of human teammates working together to compose the rover's body resonates with Morana Alač's description of the robotic "body-in-interaction." Examining a robot in a caregiving facility, she shows how the robot's "body" and its activities are negotiated through situated interaction with its interlocutors and thus "[emerge] across subjects and objects as a dynamic and interactive phenomenon" (Alač, "Moving Android," 496). I build on Alač's contribution to show how such a robotic self interactionally constituted is also contingent on the local social order.

40. Durkheim, *Elementary Forms of the Religious Life.*

41. Durkheim, *Elementary Forms of the Religious Life*, 358–59. Durkheim identifies totemism as an attribute of "primitive" societies. I align more closely with later thinkers who eschew the notion of the "primitive" and those who posit the totem as a society's way of making concrete its socially constructed relations and categories for understanding the world.

42. Collins, *Interaction Ritual Chains.* According to Collins, interaction rituals require the preconditions of physical copresence, a boundary from outsiders, mutual focus, and initial emotional impetus. In this case, however, copresence is produced through ritual performance via the virtual, technological and embodied tools that connect the team to their rovers as points of mutual focus.

43. My definition of "collective" departs from that of Clancey ("Becoming a Rover"), who articulates collective identity as a question of disciplinary formation. Here I invoke collectivity in the organizational and sociopolitical sense: for example, in contrast to hierarchical or other modes of sociopolitical experience.

44. Clancey, "Becoming a Rover"; Mirmalek, "Solar Discrepancies."

45. *Spirit* SOWG meeting, sols 1128-29, March 5, 2007.

46. End of Sol meeting, October 30, 2007.

47. Ochs, Gonzales, and Jacoby, "When I Come Down."

48. Schegloff, "Notes on a Conversational Practice"; see also Sacks, *Lectures on Conversation*, II.3 and III.8.

49. *Spirit* SOWG meeting, sols 1172–74, April 20, 2007.

50. *Spirit* SOWG meeting, sols 1172–74, April 20, 2007.

51. When I wrote this *Opportunity* had surpassed 3500 sols, but *Spirit* stopped transmitting as of sol 2210. After over a year of no contact, *Spirit* was declared dead in spring of 2011. Between March 2010 and May 2011, the Rover team avoided and even denounced the term death. They preferred to describe *Spirit* as "sleeping until we hear from her again" and held out hope for a "Lazarus situation" in which the rover would return to life. I describe *Spirit*'s death and funeral below.

52. Durkheim, *Elementary Forms of the Religious Life*, 122.

53. All Hands Meeting, June 29, 2007. The meeting lasted over two hours, and many members spoke up with comments on "the process." Several changes were implemented as a result of the meeting, including the roll/role call at the top of SOWG meetings. The meeting was judged so successful that it was instituted as a quarterly event, a new ritual in the management of the collective team. The next All Hands Meeting even included showing a video that would further cement the relation between collective teaming and safe technical practices: a NOVA special on the *Columbia* Space Shuttle disaster.

54. All Hands Meeting, June 29, 2007.

55. This point has many resonances within existing science studies literature. It reminds us again that since both imaged object and subjects are disciplined in the process of scientific seeing and representing (as in Daston and Galison, *Objectivity*, or Lynch, "Discipline and the Material Form of Images"), and since objects and subjects constitute each other, it is impossible to disentangle them (Suchman, "Subject Objects").

56. Haraway (*Modest_Witness*) invokes the "God-trick" in her discussion of the view from nowhere. This may seem to appeal to an impassive satellite's point of view on a planet. However, the orbital teams I have studied and visited do not adopt such a view. Much of the work of their mission planning and data analysis in fact relies on crafting a very situated sense of exactly where the orbiter is, since location and local conditions always affect data collection. Robert Rosenberger's postphenomenological work on the MOC camera in orbit around Mars similarly relates the importance of instrumental vision to orbital image interpretation ("Perceiving Other Planets"). Cynthia Schairer also provides a thought-provoking exploration of these issues in the case of the Rover mission ("Diffused Embodiment, Extended Visions").

Chapter Six Box

57. Interview, February 15, 2007. Ascribing personalities to robots is consistent with studies of domestic robots like Roomba or AIBO: cf. Friedman et al., "Hardware Companions?" Sung et al., "My Roomba Is Rambo."

58. *Spirit* wake, July 28, 2011; emphasis in original.

59. *Spirit* wake, July 28, 2011.

60. Melissa Rice, "Why Cuteness Matters," *Planetary Society Blog*, June 15, 2011, http://www.planetary.org/blogs/guest-blogs/3065.html.

61. *Spirit* wake, July 28, 2011; emphasis in the original.

62. Suchman, *Human-Machine Reconfigurations;* Alač, "Moving Android."

63. I am inspired here by Woolgar and Pawluch's "Ontological Gerrymandering": the ways the boundaries around the thing-which-is-to-be-explained shift in the context of different kinds of social analysis. We might witness such agential gerrymandering in other sites of human-robot encounter: members' methods that reveal how actors draw and redraw boundaries around and between different human and nonhuman participants in their sociotechnical system, construing and reconstruing locally accountable notions of agency in situ. Notably, members' agential gerrymandering here is produced through shared talk or visualization practices that *draw* the rover *as* singularly agential ("she") versus *drawing* the rover *as* collective machine ("we"). In this way, members produce multiple lay accounts of their own actor networks.

64. Interview, July 3, 2007; emphasis in the original.

65. *Spirit* wake, July 28, 2011; emphasis in original.

66. There is something of the social order in the "she" as well. Mark confessed at the wake that he thought of the rovers as "she" not only because of the "gracefulness" of their design but also "something about the vibe of the team and kind of where that personality just comes out of that makes her a *she* to me." The collective orientation of consensus is often associated with a "feminine" organizational form, as opposed to a competitive, hierarchical, "masculine" form of leadership.

Chapter Seven

1. The article in question is Brand et al., "Digital Retouching." The quotations in this paragraph are taken from my visit to Sam's laboratory the week of May 22, 2007.

2. This is, of course, a retrospective evaluation based on the introduction of a new technology. As with many actor's accounts in the history of technology, we would be wise to remain attuned to how the new tools augment and respond to existing goals and practices, while at the same time reframing them as insufficient. Historians of photography in particular will recognize the suspicious character of digital photographs as eerily reminiscent of nineteenth-century debates about photographic verisimilitude and trustworthiness; see especially Tucker, *Nature Exposed*; Tucker, "Photography as Witness, Detective, and Impostor."

3. The notion of constraints has been belabored in science studies. Constraints were the focus of a long-standing debate between sociologist of science Andy Pickering and historian of science Peter Galison about the limits of a social constructivist approach to scientific practice. Galison described constraints as "creat[ing] a problem domain, giving it shape, structure and direction" ("Contexts and Constraints," 22), implying that the natural world could act as a constraint on the growth of scientific knowledge alongside scientists' beliefs about physical laws or instruments. Pickering, in contrast, asserted that such a view can only be retrospective; material and cultural resistances encountered in "the mangle of practice" may be variously flexible, but "there is no especially informative pattern to be discovered about what changes and what does not" (*Mangle of Practice*, 207).

4. On distinctions and demarcations in scientific practice, see Burri, "Doing Distinctions"; Fyfe and Law, "Introduction: On the Invisibility of the Visual"; Gieryn, "Boundary-Work"; Lynch and Edgerton, "Aesthetics and Digital Image Processing"; Star and Griesemer, "Institutional Ecology, 'Translations,' and Boundary Objects."

5. This is consistent with work by historians of science Steven Shapin or Lorraine Daston, who focus on the modes of conduct that guarantee community-approved epistemic valid-

ity. Shapin's work on gentlemanly disinterest among members of the Royal Society draws on Goffman (*Presentation of Self in Everyday Life*) and Greenblatt (*Renaissance Self-Fashioning*) to show the importance of "gentlemanly" interactions within the early Royal Society (Shapin, *Social History of Truth*). Trustworthy actions are not limited to personal comportment: Lorraine Daston and Peter Galison (*Objectivity*) describe the "scientific self" as one crafted through local, disciplined adherence to shifting values and practices of objectivity. Doing science the right way—for that time and place—can guarantee that one is considered a trustworthy reporter in one's scientific network too. Action and interaction are therefore both essential to producing trustworthy individuals who make trusted statements about the natural world.

6. Beaulieu, "Images Are Not the (Only) Truth."

7. Galison, *Image and Logic*.

8. Rover scientist, personal conversation, February 2008.

9. The practical difficulties with replication have been explored in the sociological work of Harry Collins (*Changing Order*) and in efforts to replicate historical experiments (e.g., Sibum, "Reworking the Mechanical Value of Heat"). These scholars have shown that experimental replication demands a high degree of tacit knowledge, such that it can be impossible to replicate experimental results even given the most detailed instructions. This issue came to a head in the contested replication of Isaac Newton's prism experiments (as related in his *Opticks*), discussed by Schaffer ("Glass Works") and critiqued by Shapiro ("Gradual Acceptance of Newton's Theory of Light and Colors"). Even though replication may be nearly impossible to achieve and is rarely practiced, studies of cases such as the cold fusion debate (discussed in Collins and Pinch, *Golem*) suggest that experimental results are subject to discredit if they cannot be replicated.

10. When creating their industry-standard maps, the US Geological Survey and others rely increasingly on thermal data to provide a base map. Although the THEMIS orbital instrument measures only thermal emissions spectra, for example, the infrared and nonvisible range of light, it presents a good balance between orbital coverage and resolution on which to coregister other datasets. This is possible because different features of the Martian terrain retain and reflect heat differently; rocks are typically cooler during the day but retain their heat at night, while sand is typically hotter during the day and cools quickly at night. By comparing daytime and nighttime infrared readings, this instrument team can provide essentially a ground map of topographical features derived solely from infrared detection. Such an example further blurs the distinction between spectrometers and cameras in practice. I am grateful to site visits and interviews at the TES, THEMIS, and MiniTES headquarters at Arizona State University for discussion of these particulars.

11. Interview, Ross, June 4, 2007.

12. Notably, this took considerable work. The file formats were not compatible with ENVI, so Ross first opened them in their respective software suites and exported them to file formats that ENVI recognized. Even the most "natural" correspondence between datasets must be actively constructed.

13. Coregistering data has a political as well as a mathematical and moral dimension. Ross's attempt to coregister HRSC (camera) and OMEGA (spectrometer) data was not entirely successful for reasons he ascribed to international politics: "This HRSC data is in a fundamentally different projection that the OMEGA data is. . . . [T]he HRSC was built by the Germans and the OMEGA was built by the French, and you know how well *they* get along" (site visit, Ross, June 4, 2007; his emphasis). When he eventually presented the results at a conference in July 2007, he explained that he had to "do some gymnastics" to get the datasets to align—that is, resorting to

nonmathematical tools (field notes, Seventh International Conference on Mars, Pasadena, CA, July 9, 2007). Such political considerations are not limited to national space agencies but also arise amid the micropolitics of decision making on spacecraft teams. This heightens the significance of the decision to treat the instruments onboard the Mars Exploration Rovers as a unified suite that permits coregistration of instrumental data, a move that embeds the team's values for constrained analysis in the rovers' very construction.

14. On the meaning of models in geology, see Oreskes, "From Scaling to Simulation."

15. Interview, Sam, May 24, 2007. Problems of simulation science and reproductionism are beautifully described in Paul Edwards's discussion of climate modeling on Earth (*Vast Machine*, 263–81).

16. Observational interview, Katie, June 21, 2007.

17. Lane, "Geographers of Mars."

18. Helmreich, *Alien Ocean*. See also Lisa Messeri's "Placing Outer Space," an ethnographic study of Mars analog work across planetary science.

19. Meeting observation, July 8, 2007.

20. End of Sol meeting, May 23, 2007.

21. Team Meeting, February 13, 2008.

22. An analysis of ALH84001 published in 1996 claimed a biological origin for microscopic features in the meteorite. This is credited among planetary scientists as the reason for President Clinton's founding a separate Mars program within NASA with a mandate to launch missions to Mars every two years. This was the eventual source of funding for the Mars Exploration Rover mission.

23. Team Meeting, February 14, 2007.

24. Interview, Susan, June 18, 2007.

25. Interview, Bart, June 20, 2007.

26. Interview, June 7, 2007.

27. Team Meeting, July 7, 2007.

28. In a further twist, the computational can also be called on to provide elements of experience. When questioning whether the bright band around the rim of Victoria Crater could be ascribed to a waterline, a team member suggested something of a digital experiment. One of his colleagues had written a program that would fill any given volume with liquid: he suggested that his colleague should take the pictorial data of Victoria Crater, generate a three-dimensional volume, and "fill it with water"—digitally. This was promised to provide evidence on whether the band was consistently located the whole way around the rim, constraining the hypothesis about a water-filled Victoria Crater.

29. I am indebted here to Kohler's description of field biology's appropriation of laboratory-inspired techniques in the scientizing of field science (*Landscapes and Labscapes*).

30. Pinch, "Towards an Analysis of Scientific Observation." This notion of externality is responsible for the underdetermination of observational reports: that is, the likelihood that an interpretation may be assumed without exploring or even producing all possible available evidence.

31. Team Meeting, February 13, 2007.

32. As described above, the rover missed the selected rock and crushed another, which Sam (LTP Lead at the time) named Innocent Bystander. Only very occasionally are targets named by circumstance, as happened with Innocent Bystander or the rock called Good Question (Sam's response on the line when someone asked what to call it, which was then jokingly entered into

the software as the target name). This example was often recalled as part of *Spirit*'s personality: the rover was temperamental and did not always do what "she" was told.

33. End of Sol meeting, May 30, 2007.

34. Quotations and descriptions in this section are taken from my field notes, recordings, and photographs from my visit to Nick's lab, June 6, 2007.

35. Team Meeting, July 7, 2007.

36. Team Meeting, February 12, 2008.

37. Interview, Nick, June 6, 2007.

38. Results such as *Viking*'s inconclusive biological experiments and the controversy over the interpretation of the microstructures visible in ALH84001 have disciplined the community into extreme caution when attempting to discuss possible life on Mars. At the same time, a growing NASA Astrobiology Program presents substantial funding opportunities for scientists who present their work as related to the detection and understanding of extraterrestrial life. This tension places astrobiologists in a difficult bind with respect to their accountability to their funding agency versus their professional colleagues.

39. Interview, Nick, June 6, 2007.

40. Interview, Nick, June 6, 2007. As an ethnographer I have no doubt internalized much of the combination of excitement and discomfort I witnessed on the team during this period. I cannot pretend to write about this episode as if talk of life on Mars is just talk, with no implications for the actors involved or for their broader community. The tone of conversations I witnessed was measured and cautious, indicative of the high stakes involved in exploring such a hypothesis. After all, not only Nick's credibility but that of the entire Rover team was on the line. Nick's concern was not unfounded. An announcement in 2010 from the NASA Astrobiology Program about the discovery of arsenic-processing bacterial life forms on Earth placed a young postdoctoral scholar briefly in the limelight, only to turn into a public storm over the purported results.

41. Seventh International Conference on Mars, July 9, 2007.

42. Squyres et al., "Silica-Rich Deposits on Mars."

43. Other experts within the Mars community may have picked up the hint: shortly thereafter, a group unassociated with the Rover team published a paper in *Icarus* (the flagship journal of planetary science) under the title "A Multidisciplinary Study of Silica Sinter Deposits with Applications to Silica Identification and Detection of Fossil Life on Mars" (Preston et al.). That paper too would not assert that silica sinters could stand as "biomarkers" but only presented them as evidence for "the existence of pre-biotic conditions on Mars."

44. Interview, Sam, May 24, 2007; his emphasis.

Chapter Eight

1. The Mars Program Office funded both the rovers and *Curiosity*, but the twin rovers were funded at a Discovery-class level while *Curiosity*'s budget resembles that of the larger Flagship class missions. When *Curiosity*'s launch was delayed to November 2011 and its budget expanded accordingly, this presented significant implications for the operational budgets of the rovers, the Mars Reconnaissance Orbiter satellite, and other NASA missions.

2. End of Sol meeting, October 17, 2007.

3. End of Sol meeting, November 17, 2007; emphasis mine.

4. It may seem strange to reserve the final chapter in this book for a discussion of political factors, which are likely the first thing readers think about when they consider NASA. My pur-

pose is not to intimate that macropolitics plays only an ancillary role in Rover science, but rather to follow my actors and the scope of my laboratory field site in my analysis, concordant with my embedded perspective. I hope thus to demonstrate how a range of factors—including but not limited to macropolitical ones—play into the formulation of our visions of Mars, and how multiple ways of image construal may correspond to different intended audiences and communities.

5. See especially Biagioli, "Galileo the Emblem Maker" and *Galileo's Instruments of Credit*.

6. Bell, *Postcards from Mars*.

7. Kessler, "Spacescapes." Kessler's study of the Hubble Heritage Program shows how image processers use conventions such as the sublime, frontier photography, and European romanticism in producing Hubble space telescope images for public display. Her analysis includes framing, color palette, lighting, and other compositional elements to examine such awe-inspiring pictures as the Eagle Nebula (aka "Pillars of Creation"; see also Greenberg, "Creating the 'Pillars'"). Following Denis Cosgrove (*Social Formation and Symbolic Landscape*), she shows how spacescapes play a role in making political and social relationships appear to arise simply and naturally from this very crafted vision of outer space.

8. Previous studies of the National Aeronautics and Space Administration have noted conflicts at a variety of scales from the agency's overall bureaucratic structure to local institutional authority, and even to the mission-level management of daily resources. See Vaughan, *"Challenger" Launch Decision*; McCurdy, *Inside NASA*.

9. Macrolevel politics play a role in determining space program priorities. President George W. Bush's announcement benefited the two NASA centers that specialize in manned spaceflight, Johnson and Kennedy. These centers are in Texas and Florida, Republican strongholds where Bush family members held elected office. Conversely, President Barack Obama's cancellation of the Constellation and Shuttle programs devastated jobs at those centers but poured money and contracts into centers and private industry that focus on robotic exploration or private manned spaceflight, located in the traditionally Democratic states of California and Maryland.

10. Intermission politics can also affect local decisions. The budget crises imposed by MSL, for instance, fostered close working relations between the rovers and the MRO spacecraft, both victims of budget cuts. MSL was also frequently invoked in the competition for resources, even though many team members serve on MRO and MSL as well.

11. This is a limited budget for an organization that pays billions of dollars for mission development and allocates hundreds of millions for continuing operations. Internal to the mission, this funding cap places high demands on Rover scientists and engineers, who must continue to perform with exactitude and finesse to remain NASA's poster child mission and guarantee yearly financial support.

12. Team Meeting, July 19, 2011.

13. This can result in what appears to be a duplication of efforts but is actually an attempt to meet different institutional aims and needs. For example, rover images are processed at a facility at JPL before being posted for members to access, but they are also downloaded for processing to local servers at universities affiliated with the mission. The differences between these kinds of processing are minimal, but the institutional distinction is significant: images on JPL servers may be subject to access control, whereas universities with public mandates must provide educational opportunities for all students without discrimination based on nationality. Doubling this image work across institutions allows for essential mission work such as Pancam calibration or DEM mapping to take place.

14. I was privileged to observe a MEPAG general meeting at the Seventh International Conference on Mars in July 2007, and at the MSL landing site meetings in September 2008. Minutes and reports from the MEPAG committee are available at http://mepag.jpl.nasa.gov.

15. Interview, Simon, October 5, 2007. Other Mars scientists frequently invoke rover data as a "ground truth" to be used alongside data from orbital instruments. The MiniTES instrument, for example, is purposely similar to the orbital instruments TES and THEMIS so as to provide comparable datasets, and cooperative observations in which the surface-based spacecraft looks up at the same time that the orbital spacecraft looks down are increasingly popular since MRO's arrival at Mars.

16. Cf. Vertesi and Dourish, "Value of Data."

17. Seventh International Conference on Mars, July 10, 2007.

18. Seventh International Conference on Mars, July 12, 2007.

19. This may be the reason the Rover team believes it has never been "scooped" by other scientists on a discovery, despite its open data policy. It may also explain why it is difficult to refute published results in the broader community.

20. Interview, June 18, 2007.

21. Interview, former Mars Program Administrator, May 23, 2007.

22. I use "imagined" here not to indicate that this view is false, but rather in the sense of Anderson's *Imagined Communities*: groups united not through face-to-face interactions but through an affinity presumed between them, such as nation-states or ethnic groups.

23. Interview, former Mars Program Administrator, May 23, 2007.

24. Foucault (*Discipline and Punish*) discusses Jeremy Bentham's Panopticon as an example of prison architecture that enables the guard to observe all prisoners at any time, while no prisoner can be sure precisely when he or she is being watched. Prisoners thus discipline themselves to conform to expectations or standards of behavior even when they are not, strictly speaking, actively being watched. Just as the Panopticon serves "to induce in the inmate a state of conscious and permanent visibility that assures the automatic functioning of power" (Foucault, *Discipline and Punish*, 201), the consciousness of visibility that Rover team members exhibit also maintains power relations between the team and its patrons, the widely defined public.

25. *Opportunity* SOWG, date withheld.

26. This is a common concern. One scientist cautioned a colleague in a Team Meeting to be careful about releasing an observation in case the public "might think it's a leprechaun talking to a flamingo or something" (Team Meeting, July 6, 2007). This statement was not a joke at the public's expense, but rather reveals the keenly felt tension inherent in the management of public interpretation of images: one in which it is difficult to maintain expertise at image interpretation and manage these images' ambiguity when the public is meant to "see for themselves" and join the adventure of making discoveries on Mars.

27. Although team members usually restrain themselves from interfering with this external conversation, sometimes the temptation to say something is too hard to resist. One team member admitted to me that, after reading a heated online conversation about whether the team would take a particular image, members posted a cryptic reply suggesting that the questions would be resolved with tomorrow's downlink.

28. Cornell University astronomy colloquium, June 20, 2008.

29. Cornell University astronomy colloquium, June 20, 2008.

30. Interview, George, June 6, 2007.

31. The tension between the aesthetic and scientific principles of astronomical imaging has been well documented. In their study of the Harvard Smithsonian Observatory, Michael Lynch and Sam Edgerton describe the use of visual techniques from nonrepresentational modern art (such as pointillism and abstraction) in producing images of nonvisual astronomical phenomena such as radio pulses ("Abstract Painting and Astronomical Image Processing"). In another paper, Lynch and Edgerton discuss how members draw boundaries in their practice between the scientific and the aesthetic in crafting their images, with at least one of their informants deriding the aesthetic techniques as "a cheap way of dressing up the presentation" and "a distraction" from the science" ("Aesthetics and Digital Image Processing," 194).

32. Interview, George, June 6, 2007.

33. On the "ethical simulation" of color enhancement, see Michael Lynch, "Laboratory Space and the Technological Complex." For an actor's account of the "true colors" of the planets see Young, "What Color Is the Solar System?"

34. See http://pancam.astro.cornell.edu/pancam_instrument/true_color.html.

35. Interview, Thomas, October 5, 2007. Note that "constraints" here refers to Rover resource limitations as discussed in chapter 1, not scientific hypotheses as outlined in chapter 7.

36. The animation is online at http://www.maasdigital.com/gallery.html (accessed November 18, 2008). The rover enjoys several sunsets on its landing pad before exploring the alien terrain, and it drives off into the sunset about eight minutes, twenty-four seconds into the clip. The Walt Disney 2006 feature IMAX film *Roving Mars* relies on similar imagery in computer-produced animations to highlight the rovers' experiences as explorers of another world.

37. *Opportunity* SOWG, sol 1102, February 28, 2007.

38. *Opportunity* SOWG, sols 1063–65, January 19, 2007.

39. Ansel Adams is a frequent point of comparison in space science. The HST Hubble Heritage project includes an Ansel Adams section featuring black-and-white Hubble images (Kessler, "Spacescapes"). In an eloquent letter to the photography critic at the *New York Times*, the Viking image team's deputy leader also explained his team's work process as akin to Ansel Adams's (Viking Imaging Team deputy leader Elliott Levinthal to *New York Times* photography critic Gene Thornton, January 22, 1979, PP02.02, Elliott C. Levinthal Viking Lander Imaging Science Team Papers, 1970–80, NASA Ames History Office, NASA Ames Research Center, Moffett Field, California, 13:23).

40. My description of the picturesque and the sublime is indebted to Cosgrove, *Social Formation and Symbolic Landscape*; Burke, *Philosophical Enquiry into the Origin of Our Ideas of the Sublime and Beautiful*; Stockstad, *Art History*.

41. Lynch and Edgerton, "Aesthetics and Digital Image Processing."

42. Field notes, Astronomy 310 observations, October 2005.

43. George, personal conversation, June 6, 2007.

44. Matt Golombek et al., "PSG Letter to Project Manager [AJ Spear]" (December 9, 1994), Acquisition 022–2005, NASA Ames History Office, NASA Ames Research Center, Moffett Field, California, C2:F29.

45. This technique is often used in nature documentaries, meant to place the viewer in the action (Mittman, *Reel Nature*). Here I might also comment on Shapin and Schaffer's concept of "virtual witnessing" (*Leviathan and the Air-Pump*). Subsequent discussions of virtual witnessing have focused on what is being witnessed and how this is arranged for the observer. However, it is important to note that the stance the witness is encouraged to take can also accomplish social

work for the experimenter. Where one is asked to witness from is often as important as what one is asked to witness.

46. Interview, Ross, June 5, 2007.

47. Channel 6 News, 6:39 a.m., CBS, March 26, 2008. Source: Global Broadcast Database.

48. http://www.spacepolitics.com/2008/03/24/mars-rover-funding-cuts-will-there-be-a-backlash, Monday, March 24, 8:56 p.m.

49. http://i09.com/371700/spirit-the-mars-rover-left-to-die-before-its-time, March 25, 2008, Tuesday, 10:00 a.m. EST.

50. http://www.universetoday.com/2008/03/25/nasa-u-turn-over-mars-rover-funding, posting by MrBill March 25, 2008 at 10:46 p.m. (accessed November 18, 2008).

51. I am grateful to a conversation on the topic with Alan Stern, December 4, 2009.

52. *Spirit* SOWG, March 26, 2008.

Conclusion

1. Visiting Mark, Sarah, and Nick at JPL one afternoon, I noticed that Sarah had a bandage around one knee: a freak injury while salsa dancing, she said, incurred about the same time that *Spirit* got stuck. As Sarah stood on her good leg, favoring her bandaged knee, while discussing *Spirit* listing to one side on Mars with her wheel sunk into the sand, I recalled Jude's saying, "When the rover isn't healthy, we feel it in our bodies."

2. Field notes, January 29, 2010.

3. The team was discussing a Lazarus situation as far back as an End of Sol meeting on July 18, 2007, during a massive summer dust storm that threatened *Spirit*'s solar power acquisition.

4. Rudwick, "Emergence of a Visual Language for Geological Science."

5. Goodwin, "Professional Vision."

6. Fleck, *Genesis and Development of a Scientific Fact.*

7. Lynch, "Discipline and the Material Form of Images"; Amann and Knorr-Cetina, "Fixation of (Visual) Evidence."

8. Daston and Galison, *Objectivity.*

9. Brand et al., "Digital Retouching."

10. Lynch, "Discipline and the Material Form of Images."

11. Pinch, "Towards an Analysis of Scientific Observation."

12. On the importance of graphic traces, see Latour, "The 'Pedofil' of Boa Vista." Further, as it reveals the many kinds of practical, social, and material commitments that shape visualization, *drawing as* permits a move away from the troublesome question of what constitutes the "theory" in "theory-laden observation" and toward a praxiological orientation to visual skill, including the work of producing such images and the images' implications for future observations and interactions.

13. For example, Schaffer, "On Astronomical Drawing"; Kemp, "Temples of the Body and Temples of the Cosmos", Lynch and Edgerton, "Abstract Painting and Astronomical Image Processing."

14. For example, Schiebinger, *Mind Has No Sex?* Lisa Cartwright, *Screening the Body.*

15. Rudwick, *Scenes from Deep Time.*

16. Kaiser, *Drawing Theories Apart.*

17. Lynch, "Externalized Retina."

18. Alač, *Handling Digital Brains*; Beaulieu, "Images Are Not the (Only) Truth" and "Voxels in the Brain"; Dumit, *Picturing Personhood*; Joyce, *Magnetic Appeal.*

19. With the notable exception, of course, of Kessler, "Spacescapes"; Lynch and Edgerton, "Aesthetics and Digital Image Processing" and "Abstract Painting and Astronomical Image Processing."

20. Haraway, *Simians, Cyborgs, and Women*; Traweek, *Beamtimes and Lifetimes*; Longino, *Science as Social Knowledge*.

21. Kuhn, *Structure of Scientific Revolutions*; Hanson, *Patterns of Discovery*.

22. Shapin and Schaffer, *Leviathan and the Air-Pump*; Wittgenstein, *Philosophical Investigations*.

23. This presents fruitful historiographical considerations for the study of classic images that endure in the history of science. Novel insights arise not from *seeing* these images *as* documents, but from *seeing* them *as* crafted among moments of exchange and social relations.

24. Goodwin, "Seeing in Depth."

25. Hutchins, *Cognition in the Wild*.

26. Scott, *Seeing like a State*.

BIBLIOGRAPHY

Akrich, Madeleine. "The De-scription of Technological Objects." In *Shaping Technology/Building Society*, edited by Weibe Bijker and John Law, 205–24. Cambridge, MA: MIT Press, 1992.

Alač, Morana. *Handling Digital Brains: A Laboratory Study of Multimodal Semiotic Interaction in the Age of Computers*. Cambridge, MA: MIT Press, 2011.

———. "Moving Android: On Social Robots and Body-in-Interaction." *Social Studies of Science* 39 (2009): 491–528.

———. "Working with Brain Scans: Digital Images and Gestural Interaction in fMRI Laboratory." *Social Studies of Science* 38 (2008): 483–508.

Amman, Klaus, and Karin Knorr-Cetina. "The Fixation of (Visual) Evidence." In *Representation in Scientific Practice*, edited by Michael Lynch and Steve Woolgar, 85–121. Cambridge, MA: MIT Press, 1990.

Anderson, Benedict R. *Imagined Communities: Reflections on the Origin and Spread of Nationalism*, rev. ed. London: Verso, 1983.

Barad, Karen. *Meeting the Universe Halfway: Quantum Physics and the Entanglement of Matter and Meaning*. Durham, NC: Duke University Press, 2007.

Barker, Jane, and Hazel Downing. "Word Processing and the Transformation of Patriarchal Relations of Control in the Office." In *The Social Shaping of Technology*, edited by Donald MacKenzie and Judy Wajcman, 157–64. London: Open University Press, 1985.

Barley, Stephen. "Technology as an Occasion for Structuring: Evidence from Observations of CT Scanners and the Social Order of Radiology Departments." *Administrative Science Quarterly* 31 (1986): 78–108.

Baxandall, Michael. *Patterns of Intention: On the Historical Explanation of Pictures.* New Haven, CT: Yale University Press, 1985.

Beaulieu, Anne. "Images Are Not the (Only) Truth." *Science, Technology and Human Values* 27 (2002): 53–86.

———. "Voxels in the Brain: Neuroscience, Informatics, and Changing Notions of Objectivity." *Social Studies of Science* 31, no. 5 (2001): 635-80.

Bell, James F. III. *Postcards from Mars: The First Photographer on the Red Planet.* New York: Penguin, 2006.

Bell, James F. III, et al. "In-Flight Calibration and Performance of the Mars Exploration Rover Panoramic Camera (Pancam) Instruments." *Journal of Geophysical Research* 111 (2006): E02S03. doi:10.1029/2005JE002444.

———. "Mars Exploration Rover Athena Panoramic Camera (Pancam) Investigation." *Journal of Geophysical Research* 108 (2003): E12, 8063. doi:10.1029/2003JE002070.

Biagioli, Mario. "Galileo the Emblem Maker." *Isis* 81, no. 2 (1990): 230–58.

———. *Galileo's Instruments of Credit.* Chicago: University of Chicago Press, 2007.

Bloom, Terrie F. "Borrowed Perceptions: Harriot's Maps of the Moon." *Journal for the History of Astronomy* 9 (1978): 117–22.

Bowker, Geof, and Susan Leigh Star. *Sorting Things Out: Classification and Its Consequences.* Cambridge, MA: MIT Press, 1999.

Brand, Stewart, et al., "Digital Retouching: The End of Photography as Evidence of Anything." *Whole Earth Review* 47 (July 1998): 42–49.

Bredekamp, Horst. *Galilei der Künstler.* Berlin: Akademie Verlag, 2007.

Burke, Edmund. *A Philosophical Enquiry into the Origin of Our Ideas of the Sublime and Beautiful,* 5th ed. London: Dodsley, 1767.

Burnett, Graham. *Masters of All They Surveyed: Exploration, Geography, and a British El Dorado.* Chicago: University of Chicago Press, 2000.

Burri, Regula Valerie. "Doing Distinctions: Boundary Work and Symbolic Capital in Radiology." *Social Studies of Science* 38 (2008): 35–62.

Cartwright, Lisa. *Screening the Body: Tracing Medicine's Visual Culture.* Minneapolis: University of Minnesota Press, 1995.

Cheng, Leo, Nicole Spanovich, Alicia Vaughan, and Robert Lange. "Opposite Ends of the Spectrum: Cassini and Mars Exploration Rover Science Operations." In *Proceedings of AIAA Space-Ops 2008 Conference.* AIAA 2008–3544, May 12, 2008. AIAA doi: 10.2514/6.2008-3544.

Christensen, Phil, N. S. Gorelick, G. L. Mehall, and K. C. Murray. THEMIS Public Data Releases. Planetary Data System node, Arizona State University. http://themis-data.asu.edu.

Clancey, William. *Working on Mars: Voyages of Scientific Discovery with the Mars Exploration Rover Mission.* Cambridge, MA: MIT Press, 2012.

———. "Becoming a Rover." In *Simulation and Its Discontents,* edited by Sherry Turkle, 107–27. Cambridge: Cambridge University Press, 2009.

———. "Clear Speaking about Machines: People Are Exploring Mars, Not Robots." Paper presented at AAAI, Boston, 2008.

Collins, Harry. *Changing Order: Replication and Induction in Scientific Practice.* Chicago: University of Chicago Press, 1992.

Collins, Harry, and Robert Evans. "The Third Wave of Science Studies: Studies of Expertise and Experience." *Social Studies of Science* 32, no. 2 (2002): 235–96.

Collins, Harry, and Trevor Pinch, *The Golem: What Everyone Should Know about Science.* Cambridge: Canto, 1993.

Collins, Randall. *Interaction Ritual Chains.* Princeton, NJ: Princeton University Press, 2004.

Coopmans, Catelijne. "'Face Value': New Medical Imaging Software in Commercial View." *Social Studies of Science* 41 (2010): 155–76.

Cosgrove, Denis. *Geography and Vision: Seeing, Imagining and Representing the World.* London: Taurus, 2008.

———. *Social Formation and Symbolic Landscape.* Milwaukee: University of Wisconsin Press, 2009.

Coulter, Jeff, and E. D. Parsons. "The Praxiology of Perception: Visual Orientations and Practical Action." *Inquiry* 33 (1991): 251–72.

Daston, Lorraine, and Peter Galison. "The Image of Objectivity." *Representations* 40 (1992): 81–128.

———. *Objectivity.* Cambridge: Zone Books, 2007.

Delbourgo, James, and Nicholas Dew, eds. *Science and Empire in the Atlantic World.* New York: Routledge, 2007.

DiSalvo, Carl, and Francine Gemperie, "From Seduction to Fulfillment: The Use of Anthropomorphic Form in Design." In *Proceedings of DPPI 2003,* 67–72. New York: ACM Press, 2003.

DiSalvo, Carl, et al. "All Robots Are Not Created Equal: The Design and Perception of Humanoid Robot Heads." In *Proceedings of the 2002 ACM Conference on Designing Interactive Systems,* 321–26. New York: ACM Press, 2002.

Doing, Park. "'Lab Hands' and the 'Scarlet O': Epistemic Politics and (Scientific) Labor." *Social Studies of Science* 34 (2004): 299–323.

Downer, John. "When the Chick Hits the Fan: Representativeness and Reproducibility in Technological Testing." *Social Studies of Science* 31, no. 1 (2007): 7–26.

Dumit, Joseph. *Picturing Personhood: Brain Scans and Biomedical Identity.* Princeton, NJ: Princeton University Press, 2003.

Durkheim, Émile. *The Division of Labor in Society.* 1893. Translated by W. D. Halls, 1933. Reprint, New York: Simon and Schuster, 1997.

———. *The Elementary Forms of the Religious Life.* 1912. Reprint, Paris, 1915.

Edgerton, Samuel. "Galileo, Florentine 'Disegno,' and the 'Strange Spottednesse' of the Moon." *Art Journal* 44 (1984): 225–32.

Edwards, Paul. *A Vast Machine: Computer Models, Climate Data, and the Politics of Global Warming.* Cambridge, MA: MIT Press, 2011.

Fleck, Ludwig. *Genesis and Development of a Scientific Fact.* Edited by Thaddeus Trenn and Robert K. Merton. Translated by Fred Bradley and Thaddeus Trenn, 1935. Reprint, Chicago: University of Chicago Press, 1979.

Foucault, Michel. *Discipline and Punish: The Birth of the Prison.* Translated by Alan Sheridan. New York: Vintage Books, 1977.

Franklin, Alan. "Calibration." *Perspectives on Science* 5 (1997): 31–80.

Friedman, Batya, et al. "Hardware Companions? What Online AIBO Discussion Forums Reveal about the Human-Robotic Relationship." In *Proceedings of the 2003 ACM Conference on Human Factors in Computing Systems,* 273–80. New York: ACM Press, 2003.

Fyfe, Gordon, and John Law. "Introduction: On the Invisibility of the Visual." In *Picturing Power: Visual Depiction and Social Relations,* edited by Gordon Fyfe and John Law, 1–14. London: Routledge, 1988.

Galison, Peter. "Contexts and Constraints." In *Scientific Practice: Theories and Stories of Doing Physics*, edited by Jed Buchwald, 13–41. Chicago: Chicago University Press, 1995.

———. *Image and Logic*. Chicago: University of Chicago Press, 1997.

———. "Judgment against Objectivity." In *Picturing Science Producing Art*, edited by Carolyn Jones and Peter Galison, 327–59. London: Routledge, 1998.

Garfinkel, Harold. *Ethnomethodology's Program: Working Out Durkheim's Aphorism*. Edited by Anne Rawls. Lanham, MD: Rowman and Littlefield, 2002.

———. *Studies in Ethnomethodology*. Englewood Cliffs, NJ: Prentice-Hall, 1967.

Gerovitch, Slava. "'New Soviet Man' Inside Machine: Human Engineering, Spacecraft Design and the Construction of Communism. *Osiris* 22 (2007): 135–57.

Gieryn, Thomas. "Boundary-Work and the Demarcation of Science from Non-science: Strains and Interests in Professional Ideologies of Scientists." *American Sociological Review* 48 (1983): 781–95.

Goffman, Erving. *Interaction Ritual*. Chicago: Aldine, 1967.

———. *The Presentation of Self in Everyday Life*. New York: Doubleday, 1959.

Goodwin, Charles. "Professional Vision." *American Anthropologist* 96 (1996): 606–33.

———. "Seeing in Depth." *Social Studies of Science* 35 (1995): 237–74.

Greenberg, Joshua. "Creating the 'Pillars': Multiple Meanings of a Hubble Image." *Public Understanding of Science* 13 (2004): 83–95.

Greenblatt, Stephen. *Renaissance Self-Fashioning: From More to Shakespeare*. Chicago: University of Chicago Press, 1983.

Gusterson, Hugh. *Nuclear Rites: A Weapons Laboratory at the End of the Cold War*. Berkeley: University of California Press, 1998.

Hacking, Ian. *Representing and Intervening: Introductory Topics in the Philosophy of Natural Science*. Cambridge: Cambridge University Press, 1983.

Hanson, Norwood Russell. *Patterns of Discovery*. Cambridge: Cambridge University Press, 1958.

Haraway, Donna. *Modest_Witness@Second_Millennium.FemaleMan©_Meets_OncoMouse™*. New York: Routledge, 1997.

———. *Simians, Cyborgs, and Women: The Reinvention of Nature*. New York: Routledge, 1991.

———. *When Species Meet*. Minneapolis: University of Minnesota Press, 2007.

Harley, J. B. *The New Nature of Maps: Essays in the History of Cartography*. Baltimore, MD: Johns Hopkins University Press, 2001.

Helmreich, Stephan. *Alien Ocean: Anthropological Voyages in Microbial Seas*. Berkeley: University of California Press, 2009.

Hobbes, Thomas. *Leviathan*. London, 1651.

Hooke, Robert. *Micrographia*. London, 1665.

Howell, Steve B. *Handbook of CCD Astronomy*. 2nd ed. Cambridge Observing Handbooks for Research Astronomers series. Cambridge: Cambridge University Press, 2006.

Hubbard, Scott, Farouz Naderi, and J. Garvin. "Following the Water: The New Program for Mars Exploration." Presented at Fifty-Seventh International Astronautical Congress. Toulouse, France, 2001. doi: IAF-01-Q.3.a.01.

Hutchins, Edwin. *Cognition in the Wild*. Cambridge, MA: MIT Press, 1995.

Idhe, Don. *Postphenomenology and Technoscience*. Albany: SUNY Press, 2009.

Jastrow, Joseph. "The Mind's Eye." *Appleton's Popular Science Monthly* 54 (1899): 299–312.

Joyce, Kelly. "From Numbers to Pictures: The Development of Magnetic Resonance Imaging and the Visual Turn in Medicine." *Science as Culture* 15, no. 1 (2006): 1–22.

———. *Magnetic Appeal: MRI and the Myth of Transparency*. Ithaca, NY: Cornell University Press, 2008.

Kaiser, David. *Drawing Theories Apart: The Dispersion of Feynman Diagrams in Postwar Physics*. Chicago: University of Chicago Press, 2005.

———, ed. *Pedagogy and the Practice of Science: Historical and Contemporary Perspectives*. Cambridge, MA: MIT Press, 2005.

Kaminski, Amy P. "Faster, Better, Cheaper: A Sociotechnical Perspective on Programmatic Choice, Success, and Failure in NASA's Solar System Exploration Program." In *Exploring the Solar System: The History and Science of Planetary Probes*, 77–101. New York: Palgrave Macmillan, 2013.

Kemp, Martin. *Spatial Visions: The Nature Book of Art and Science*. Oxford: Oxford University Press, 2000.

———. "Taking It on Trust: Form and Meaning in Naturalistic Representation." *Archives of Natural History* 17 (1990): 127–88.

———. "Temples of the Body and Temples of the Cosmos: Vision and Visualisation in the Vesalian and Copernican Revolutions." In *Picturing Knowledge: Historical and Philosophical Problems concerning the Use of Art in Science*, edited by Brian Baigrie, 40–85. Toronto: University of Toronto Press, 1996.

Kessler, Elizabeth. "Spacescapes: Romantic Aesthetics and the Hubble Space Telescope Images." PhD diss., University of Chicago, 2006.

Kinch, Kjartan, Jascha Sohl-Dickstein, James F. Bell, Jeff Johnson, W. Goetz, and Geoff Landis. "Dust Deposition on the Mars Exploration Rover Panoramic Camera (Pancam) Calibration Targets." *Journal of Geophysical Research* 112 (2007): E06S03. doi: 10.1029/2006JE002807.

Knorr-Cetina, Karin. *Epistemic Cultures: How the Sciences Make Knowledge*. Princeton, NJ: Princeton University Press, 1999.

Knorr-Cetina, Karin, and Klaus Amman. "Image Dissection in Natural Scientific Inquiry." *Science, Technology and Human Values* 15 (1990): 259–83.

Knorr-Cetina, Karin, and Urs Bruegger. "The Market as an Object of Attachment: Exploring Postsocial Relations in Financial Markets." *Canadian Journal of Sociology* 25 (2000): 141–68.

Kohler, Robert. *Landscapes and Labscapes: Exploring the Lab-Field Border in Biology*. Chicago: University of Chicago Press, 2002.

Krygier, John, and Denis Wood. "Ce n'est pas le monde (This is not the world)." In *Rethinking Maps: New Frontiers in Cartographic Theory*, edited by Martin Dodge, Rob Kitchin, and Chris Perkins, 189–219. New York: Routledge, 2009.

Kuhn, Thomas. *The Copernican Revolution*. Cambridge, MA: Harvard University Press, 1957.

———. *The Structure of Scientific Revolutions*. 2nd ed. Chicago: University of Chicago Press, 1962.

Landis, Geoff. "Some MER Terminology: A Vocabulary Compiled by Geoff Landis." Manuscript in the author's possession.

Lane, Maria. "Geographers of Mars: Cartographic Inscription and Exploration Narrative in Late Victorian Representations of the Red Planet." *Isis* 96 (2005): 477–506.

———. *Geographies of Mars: Seeing and Knowing the Red Planet*. Chicago: University of Chicago Press, 2010.

Latour, Bruno. *Aramis, or The Love of Technology*. Cambridge, MA: Harvard University Press, 1996.

———. "Drawing Things Together." In *Representation in Scientific Practice*, edited by Michael Lynch and Steve Woolgar, 19–68. Cambridge, MA: MIT Press, 1990.

———. "The 'Pedofil' of Boa Vista: A Photo-philosophical Montage." *Common Knowledge* 4 (1995): 144–87.

———. *Science in Action: How to Follow Scientists and Engineers through Society*. Cambridge, MA: Harvard University Press, 1987.

———. "Visualization and Cognition: Thinking with Eyes and Hands." *Knowledge and Society* 6 (1986): 1–40.

Latour, Bruno, and Steve Woolgar. *Laboratory Life: The Construction of Scientific Facts*. Princeton, NJ: Princeton University Press, 1979.

Law, John, and Michael Lynch. "Lists, Field Guides, and the Descriptive Organization of Seeing: Birdwatching as an Exemplary Observational Activity." In *Representation in Scientific Practice*, edited by Michael Lynch and Steve Woolgar, 267–99. Cambridge, MA: MIT Press, 1990.

Levinthal, Elliott C. Viking Lander Imaging Science Team Papers, 1970–80. NASA Ames History Office and Archives. NASA Ames Research Center, Moffett Field, Mountain View, California.

Lewis, Kevin, Oded Aharonson, John Grotzinger, Steven Squyres, James F. Bell, Larry Crumpler, and Mariek Schmidt. "Structure and Stratigraphy of Home Plate from the *Spirit* Mars Exploration Rover." *Journal of Geophysical Research* 113 (2008): E12S36. doi:10.1029/2007JE003025.

Longino, Helen. *Science as Social Knowledge: Values and Objectivity in Scientific Inquiry*. Princeton, NJ: Princeton University Press, 1990.

Lowell, Percival. *Mars*. 1895. Reprint, Waterbury, CT: Brohan Press, 2002.

Lynch, Michael. *Art and Artifact in Laboratory Science*. London: Routledge and Kegan Paul, 1985.

———. "Discipline and the Material Form of Images: An Analysis of Scientific Visibility." *Social Studies of Science* 15 (1985): 37–66.

———. "The Externalized Retina: Selection and Mathematization in the Visual Documentation of Objects in the Life Sciences." In *Representation and Scientific Practice*, edited by Michael Lynch and Steve Woolgar, 153–86. Cambridge, MA: MIT Press, 1990.

———. "Laboratory Space and the Technological Complex: An Investigation of Topical Contextures." *Science in Context* 4 (1991): 51–78.

———. "Science in the Age of Mechanical Reproduction: Moral and Epistemic Relations between Diagrams and Photographs." *Biology and Philosophy* 6 (1991): 205–26.

———. *Scientific Practice and Ordinary Action: Ethnomethodology and Social Studies of Science*. Cambridge: Cambridge University Press, 1993.

Lynch, Michael, and Samuel Y. Edgerton. "Abstract Painting and Astronomical Image Processing." In *The Elusive Synthesis: Aesthetics and Science*, edited by F. Tauber, 103–24. Dordrecht: Kluwer Academic Publishers, 1996.

———. "Aesthetics and Digital Image Processing: Representational Craft in Contemporary Astronomy." In *Picturing Power: Visual Depictions and Social Relations*, edited by Gordon Fyfe and John Law, 184–220. Sociological Review Monograph. London: Routledge, 1988.

Maki, Justin, et al. "Operation and Performance of the Mars Exploration Rover Imaging System on the Martian Surface." In *IEEE: Systems, Man and Cybernetics*, 930–36. Waikoloa, Hawaii: IEEE, 2005.

Marcus, George. "Ethnography in/of the World System: The Emergence of Multi-sited Ethnography." *Annual Review of Anthropology* 24 (1995): 95-117.

McCurdy, Howard. *Faster, Better, Cheaper: Low-Cost Innovation in the US Space Program*. Baltimore, MD: Johns Hopkins University Press, 2001.

———. *Inside NASA: High Technology and Organizational Change in the US Space Program.* Baltimore, MD: Johns Hopkins University Press, 1993.

Merleau-Ponty, Maurice. *The Phenomenology of Perception.* 1944. Reprint, London: Routledge and Kegan Paul, 1962.

Messeri, Lisa. "Placing Outer Space: An Earthly Ethnography of Other Worlds." PhD diss., Massachusetts Institute of Technology, 2011.

Meyer, John, and Brian Rowan. "Institutionalized Organizations: Formal Structure as Myth and Ceremony." *American Journal of Sociology* 83 (1977): 340–63.

Mialet, Hélène. *Hawking Incorporated: Steven Hawking and the Anthropology of a Knowing Subject.* Chicago: University of Chicago Press, 2012.

Miller, Victor C., and Calvin F Miller. *Photogeology.* New York: McGraw-Hill, 1961.

Mindell, David. *Digital Apollo: Human and Machine in Spaceflight.* Cambridge, MA: MIT Press, 2011.

Mirmalek, Zara. "Solar Discrepancies: Mars Exploration and the Curious Problem of Interplanetary Time." PhD diss., University of California at San Diego, 2008.

———. "Working Time on Mars." *KronoScope* 8, no. 2 (2008): 159–78.

Mittman, Gregg. *Reel Nature: America's Romance with Wildlife Film.* Seattle: University of Washington Press, 1999.

Mol, Annemarie. *The Body Multiple: Ontology in Medical Practice.* Durham, NC: Duke University Press, 2003.

Mulkay, Michael, and G. Nigel Gilbert. "Joking Apart: Some Recommendations concerning the Analysis of Scientific Culture." *Social Studies of Science* 12 (1982): 585–613.

Myers, Natasha. "Molecular Embodiments and the Body-Work of Modeling in Protein Crystallography." *Social Studies of Science* 38 (2008): 163–99.

NASA Ames History Office and Archives. Acquisition 022–2005. NASA Ames Research Center, Moffett Field, Mountain View, California.

Ochs, Elinor, Patrick Gonzales, and Sally Jacoby. "'When I Come Down I'm in the Domain State': Grammar and Graphic Representation in the Interpretive Activity of Physicists." In *Interaction and Grammar*, edited by Emmanuel Schegloff and Sandra A. Thompson, 328–270. New York: Cambridge University Press, 1996.

Oreskes, Naomi. "From Scaling to Simulation: Changing Meanings and Ambitions of Models in Geology." In *Science without Laws: Model Systems, Cases, Exemplary Narratives*, edited by Angela Creager, Elizabeth Lunbeck, and M. Norton Wise, 93–124. Durham, NC: Duke University Press, 2007.

Orr, Julian. *Talking about Machines: An Ethnography of a Modern Job.* Ithaca, NY: Cornell University Press, 1996.

Perlow, Leslie, and Nelson Repenning. "The Dynamics of Silencing Conflict." *Research in Organizational Behavior* 29 (2009): 195–223.

Perrow, Charles. *Normal Accidents: Living with High-Risk Technologies.* New York: Basic Books, 1984.

Pickering, Andrew. *The Mangle of Practice: Time, Agency and Science.* Chicago: University of Chicago Press, 1995.

Pinch, Trevor. "Testing, One, Two, Three, Testing: Towards a Sociology of Testing." *Science, Technology and Human Values* 18 (1993): 25–41.

———. "Towards an Analysis of Scientific Observation: The Externality and Evidential Significance of Observation Reports in Physics." *Social Studies of Science* 15 (1985): 167–87.

Polanyi, Michael. *The Tacit Dimension.* Gloucester, MA: Smith, 1966.

Polletta, Francesca. *Freedom Is an Endless Meeting*. Chicago: University of Chicago Press, 2002.

Powell, Mark, et al. "Targeting and Localization for Mars Rover Operations." In *Proceedings of IEEE Conference on Information Reuse and Integration*, 23–27. Waikola, HI, 2006.

Prentice, Rachel. *Bodies of Information: An Ethnography of Anatomy and Surgery Education*. Durham, NC: Duke University Press, 2012.

———. "Drilling Surgeons: The Social Lessons of Embodied Surgical Learning." *Science, Technology and Human Values* 32, no. 5 (2007): 534–53.

Preston, L. J., et al. "A Multidisciplinary Study of Silica Sinter Deposits with Applications to Silica Identification and Detection of Fossil Life on Mars." *Icarus* 198, no. 2 (2008): 331–50.

Pumfrey, Stephen. "Harriot's Maps of the Moon: New Interpretations." *Notes and Records of the Royal Society* 63 (2009): 163–68.

Radder, Hans. *The World Observed/The World Conceived*. Pittsburgh: University of Pittsburgh Press, 2006.

Rosenberger, Robert. "Perceiving Other Planets: Bodily Experience, Interpretation, and the Mars Orbiter Camera." *Human Studies* 31 (2008): 63–75.

Rudwick, Martin. "The Emergence of a Visual Language for Geological Science, 1760–1840." *History of Science* 14 (1976): 148–95.

———. *Scenes from Deep Time: Early Pictorial Representations of the Prehistoric World*. Chicago: University of Chicago Press, 1992.

Sacks, Harvey. *Harvey Sacks: Lectures on Conversation*. Edited by Gail Jefferson. Oxford: Blackwell, 1992.

Salonius, Annalisa. "Social Organization of Work in Academic Labs in the Biomedical Sciences in Canada: Socio-historical Dynamics and the Influence of Research Funding." *Social Studies of Science, forthcoming*.

Schaffer, Simon. "Astronomers Mark Time: Discipline and the Personal Equation." *Science in Context* 2 (1988): 115–45.

———. "Glass Works: Newton's Prisms and the Uses of Experiment." In *The Uses of Experiment: Studies in the Natural Sciences*, edited by David Gooding, Trevor Pinch, and Simon Schaffer, 67–104. Cambridge: Cambridge University Press, 1989.

———. "On Astronomical Drawing." In *Picturing Science Producing Art*, edited by Carolyn Jones and Peter Galison, 441–74. London: Routledge, 1998.

———. "'On Seeing Me Write': Inscription Devices in the South Seas." *Representations* 97 (2007): 90–122.

Schairer, Cynthia. "Diffused Embodiment, Extended Visions: The Prosthetics of Martian Geology; Are the Mars Rovers Agents of Situated Knowledge?" Paper presented at the Annual Meeting of the Society for the Social Studies of Science, Vancouver, BC, November 2, 2006.

Schegloff, Emmanuel. "Notes on a Conversational Practice: Formulating Place." In *Studies in Social Interaction*, edited by David Sudnow, 75–119. New York: Macmillan, 1972.

Schiebinger, Londa. *The Mind Has No Sex? Women in the Origins of Modern Science*. Cambridge, MA: Harvard University Press, 1991.

Scott, James. *Seeing like a State: How Certain Schemes to Improve the Human Condition Have Failed*. New Haven, CT: Yale University Press, 1998.

Shapin, Steven. "The Invisible Technician." *American Scientist* 77 (1989): 554–63.

———. *A Social History of Truth: Civility and Science in Seventeenth-Century England*. Chicago: University of Chicago Press, 1994.

Shapin, Steven, and Simon Schaffer. *Leviathan and the Air-Pump: Hobbes, Boyle, and the Experimental Life*. Princeton, NJ: Princeton University Press, 1985.

Shapiro, Alan. "The Gradual Acceptance of Newton's Theory of Light and Colors, 1672–1727." *Perspectives of Sciences* 4 (1996): 59–140.

Shrum, Wes, Joel Genuth, and Ivan Chompalov. *Structures of Scientific Collaboration*. Cambridge, MA: MIT Press, 2007.

Sibum, Otto. "Reworking the Mechanical Value of Heat: Instruments of Precision and Gestures of Accuracy in Early Victorian England." *Studies in History and Philosophy of Science* 26 (1995): 73–106.

Squyres, Steven. *Roving Mars: "Spirit," "Opportunity," and the Exploration of the Red Planet*. New York: Hyperion, 2005.

Squyres, Steven, et al. "Detection of Silica-Rich Deposits on Mars." *Science* 320, no. 5879 (2008): 1063–67.

Star, Susan Leigh, and James Griesemer. "Institutional Ecology, 'Translations,' and Boundary Objects: Amateurs and Professionals in Berkeley's Museum of Vertebrate Zoology, 1907–39." In *The Science Studies Reader*, edited by Mario Biagioli, 505–24. New York: Routledge, 1999.

Star, Susan Leigh, and Martha Lampland. *Standards and Their Stories*. Ithaca, NY: Cornell University Press, 2008.

Stockstad, Marilyn. *Art History*. Vol. 2. New York: Prentice Hall and Harry Abrams, 1995.

Suchman, Lucy. *Human-Machine Reconfigurations: Plans and Situated Actions*. 2nd ed. Cambridge: Cambridge University Press, 2007.

———. "Subject Objects." *Feminist Theory* 12 (2011): 119–45.

Sung, J.-Y., et al. "'My Roomba Is Rambo': Intimate Home Appliances." In *Proceedings of the 2007 ACM Conference on Ubiquitous Computing Systems (Ubicomp)*, 145–62. New York: ACM Press, 2007.

Tollinger, Irene, Christian D. Schunn, and Alonso H. Vera. "What Changes When a Large Team Becomes More Expert? Analyses of Speedup in the Mars Exploration Rovers Science Planning Process." In *Proceedings of the 2006 Cognitive Science Society Conference, Vancouver, BC* (2006). http://www.cogsci.rpi.edu/CSJarchive/Proceedings/2006/docs/p840.pdf. Accessed November 10, 2008.

Traweek, Sharon. *Beamtimes and Lifetimes: The World of High Energy Physicists*. Cambridge, MA: Harvard University Press, 1988.

Tucker, Jennifer. *Nature Exposed: Photography as Eyewitness in Victorian Science*. Baltimore, MD: Johns Hopkins University Press, 2005.

———. "Photography as Witness, Detective, and Impostor: Visual Representation in Victorian Science." In *Victorian Science in Context*, edited by Bernie Lightman, 387–408. Chicago: University of Chicago Press, 1997.

Turner, Fred. "Where the Counterculture Met the New Economy: The WELL and the Origins of Virtual Community." *Technology and Culture* 46 (2005). 485–512.

Vaughan, Diane. *The "Challenger" Launch Decision: Risky Technology, Culture and Deviance at NASA*. Chicago: University of Chicago Press, 1996.

Vertesi, Janet. "Mind the Gap: the London Underground Map and Users' Representations of Urban Space." *Social Studies of Science* 38 (2008): 1–32.

———. "Picturing the Moon: Hevelius' and Riccioli's Visual Debate." *Studies in History and Philosophy of Science* 38 (2007): 401–21.

Vertesi, Janet, and Paul Dourish, "The Value of Data: Considering the Context of Production in Data Economies." In *Proceedings of the 2011 ACM Conference on Computer-Supported Cooperative Work*, 533–42. New York: ACM Press, 2011.

Wales, Roxana, Deborah Bass, and Valerie Shalin. "Requesting Distant Robotic Action: An Ontology of Work, Naming and Action Identification for Planning on the Mars Exploration Rover Mission." *Journal for the Association of Information Systems* 8, no. 2 (2007): 75–104.

Wang, Alian, and Z. C. Ling. "Ferric Sulfates on Mars—a Combined Mission Data Analysis of Salty Soils at Gusev Crater and Laboratory Experimental Investigations." *Journal of Geophysical Research* 116 (2011): E00F17. doi:10.1029/2010JE003665.

Wang, Alian, et al. "Light-Toned Salty Soils and Co-existing Si-Rich Species Discovered by the Mars Exploration Rover *Spirit* in Columbia Hills." *Journal of Geophysical Research* 113 (2008): E12S40. doi:10.1029/2008JE003126.

Weber, Max. *Theory of Social and Economic Organization*. Edited by Talcott Parsons. New York: Free Press, 1947.

Wilhelms, Don. "Geologic Mapping." In *Planetary Mapping*, edited by Ron Greeley and Raymond M. Batson, 208–60. Cambridge: Cambridge University Press, 2007.

Winichukl, Thongchai. *Siam Mapped: The History of a Geo-body of a Nation*. Honolulu: University of Hawaii Press, 1994.

Winkler, Mary, and Albert Van Helden. "Representing the Heavens: Galileo and Visual Astronomy." *Isis* 83 (1992): 195–217.

Wise, Norton, ed. *The Values of Precision*. Princeton, NJ: Princeton University Press, 1995.

Wittgenstein, Ludwig. *Philosophical Investigations*. 1953. Translated by G. E. M. Anscombe. Reprint, Oxford: Blackwell Press, 2001.

Wood, Denis. *The Power of Maps*. London: Guilford Press, 1992.

Woolgar, S., and D. Pawluch. "Ontological Gerrymandering: The Anatomy of Social Problems Explanations." *Social Problems* 32 (1985): 214–27.

Young, A. T. "What Color Is the Solar System?" *Sky and Telescope* 5 (May 1985): 399.

Zabusky, Stacia. *Launching Europe: An Ethnography of Cooperation in European Space Science*. Princeton, NJ: Princeton University Press, 1995.

INDEX

actor-network theory, 96–98, 101, 106, 178; annotation and, 141 (*see also* annotation); calibration and, 71, 274n44 (*see also* calibration); constraints and, 192, 193, 208, 214, 288n3 (*see also* constraints); "drawing as," 95, 103, 278n39, 288n63 (*see also* "drawing as"); human/ robots, 176, 265n4, 285n10, 288n63 (*see also* human/machine interfaces); language and, 19–20, 22, 264n37, 268n32, 274n40 (*see also specific topics*); politics and, 161, 221, 263n23, 282n3; science/ operations, 28, 262n18, 264n37; scripts and, 273n22; social order and, 263n23, 265n6, 274n40, 283n20 (*see also* social order); topologies, 265n40, 278n37, 278n39, 286n41, 288n63. *See also specific topics*

Adams, A., 227, 228, 229, 269n55, 271n79, 294n39

Alač, M., 175, 189, 261n8, 263n31, 272n6, 275n2, 285n10, 286n39

Amman, K., 9, 114, 262n9, 268n18

anaglyphs, 123, 129, 169, 169f, 170, 281n40, 284n9

annotation, 22, 283n19; attention and, 112–16; collectivity and, 22, 161 (*see also* collectivity); color and, 95, 116, 121, 121f (*see also* color); coordination and, 112–16, 160, 278n37; disagreement and, 160–61; "drawing as" (*see* "drawing as"); geology and, 110, 112, 121 (*see also* geology); Greek letters in, 114–15, 115f; information and, 95, 107; as interpretation, 95, 112–16, 147–48; intervention and, 106; mapping and, 112–16, 131, 137 (*see also* mapping); naming and, 120, 121f (*see also* naming); proxies and, 115; representation and, 106, 116; social order and, 147, 154, 161 (*see also* social order); software for, 114 (*see also* software); trafficability and, 111–31, 117f, 119f, 137, 139f (*see also* trafficability). *See also specific sites and topics*